Bob's House

Understanding the ACCA Residential HVAC System Design Process

Version 1.20
2016

To comment on the content of this document or for information pertaining to technical content, respond to:

Wes Davis
ACCA
1330 Braddock Place, Suite 350
Alexandria, VA 22314
703-575-4477

This fictional case study was written to assist you in understanding the ACCA residential heating, ventilation, and air conditioning (HVAC) design process. Residential load calculations, HVAC equipment selection, HVAC duct sizing, and air distribution testing, adjusting, and balancing are all discussed in a simple fashion suitable for the lay person and expert alike.

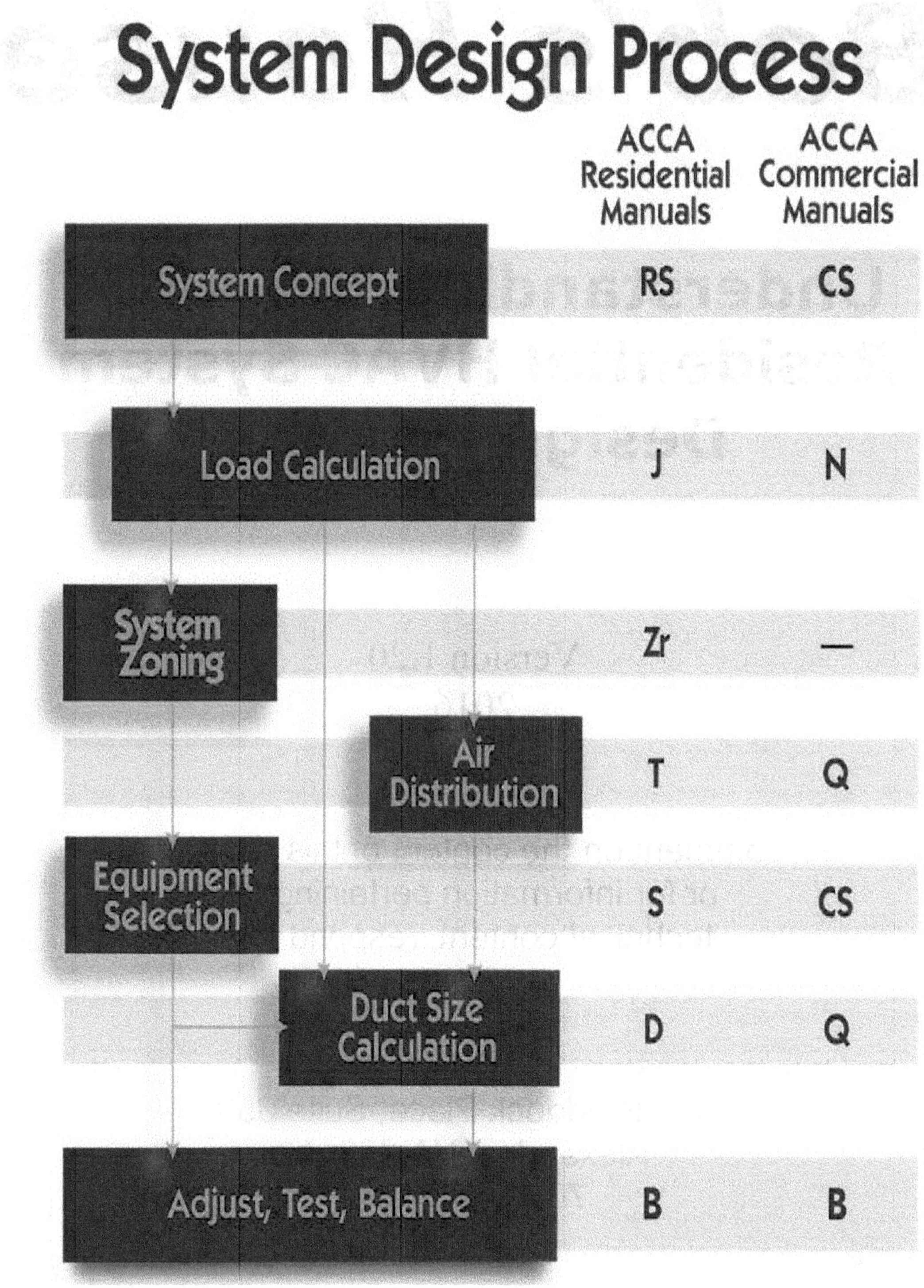

OVERVIEW

The ACCA residential heating and cooling system design process is a proven method that has been successfully used for residential applications in all climate zones, and is now becoming recognized and enforced by the majority of national and local building codes[1]. The ACCA design process has been used in countless homes with great success. However, there are still many people who would benefit from a better understanding of: (a) the residential HVAC design process, (b) how different parts of the design process interrelate, and (c) how to apply information from the calculations for the best results. This case study is meant to provide that clarification. Whether you are reviewing a permit application or a proposal from a residential heating, ventilation, and air conditioning (HVAC) contractor, you need to understand the underlying concepts. This guided tour through "Bob's House" will provide an overview of the key elements, how they interrelate, and the design sequence for:

1. Calculating a home's heating and cooling requirements.
2. Selecting the equipment to meet those requirements.
3. Sizing the duct system to deliver the correct amount of conditioned air.
4. Choosing the correct grilles and registers for optimum mixing of the conditioned air for maximum comfort.

Most contractors have heard of, or are familiar with, the first step in the design process: The ANSI-approved ACCA *Manual J®* Eighth Edition (*Manual J8*) procedure for calculating the heat loss / heat gain for a home. Some believe *Manual J8* is all that is necessary for satisfactory design. However, as you will see, this important calculation is only the first important step in the journey to achieving good heating and cooling system design. This case study discusses the residential design process from start to finish, using Bob's house as an example. Although it provides a comprehensive solution for this particular example, it does not cover every scenario, or investigate every possible solution.

[1] Enforced by: 2015 International Residential Code™: (*Manuals J, D & S*)
2015 International Energy Conservation Code™ (*Manuals J & S*).
2015 Uniform Mechanical Code™: (*Manuals J, D, B, & S*).

ACKNOWLEDGEMENTS

This update and the original document have received invaluable comments and helpful input from numerous knowledgeable experts. Chief among these were:

Jack Bartell (Virginia Air Distributors; Richmond, VA)

Ron Bladen (Air Conditioning Contractors of America; Arlington, VA)

Chris Colditz (Laco Mechanical Services; Elk Grove Village, IL)

Rob Falke (National Comfort Institute, Cleveland, OH)

Russ Goyen (Chief Plumbing and Mechanical Inspector; Idaho Falls, ID)

Jim Herritage (Energy Auditors; Mt. Pleasant, SC)

Glenn Hourahan, P.E. (Air Conditioning Contractors of America; Arlington, VA)

Dave Hutchins (Bay Area Air Conditioning; Crystal River FL)

Keith Jones (Vertical Turbine Specialists; Lubbock, TX)

Warren Lupson (Lupson and Associates; Silver Springs, MD)

Darren B. Meyers, P.E. (International Energy Conservation Consultants, LLC)

Tina Neace (Florida Air Design Inc; Umatilla, FL)

John Parker (Consultant; Clanton, AL)

Kelly Parker (Guaranteed Watt Saver Services; Oklahoma City, OK)

Donald Prather (Air Conditioning Contractors of America; Arlington, VA)

Hank Rutkowski, P.E. (H.T.R. Consulting; Dover, OH)

John Sedine (Engineered Heating and Cooling; Walker, MI)

Dick Shaw (Air Conditioning Contractors of America; Arlington, VA)

Bruce Silverman (Airite Air Conditioning; Tampa, FL)

Bill Smith (Elite; College Station, TX)

Dave Swett (HVAC Training Center, Omaha Public Power; Omaha, NE)

Chandler von Schrader (Environmental Protection Agency/Energy Star; Washington, D.C.)

Pete Van Lancker (Rheem Air Conditioning, Fort Smith; AR)

Eric Werling (Environmental Protection Agency/Indoor Environments Division; Washington, D.C.)

Bill Wright (Wrightsoft; Boston, MA)

Laurie Hourahan, and Ron Bladen provided editorial review, all of which is appreciated and acknowledged.

TABLE OF CONTENTS

CHAPTER I: *MANUAL J*: UNDERSTANDING A HOME'S HEATING / COOLING REQUIREMENTS

Meet Bob

Most people want a safe, secure, and comfortable home for their family. Inherent within the comfort of a home is the heating and cooling system. Some HVAC[2] contractors use the industry standard (code-required) design procedures outlined in this book. Unfortunately, many do not. They guess at equipment and duct sizing requirements by using rules-of-thumb or unproven shortcuts they may have learned from others while installing duct. This book walks through the correct, industry-adopted, design procedures for a heating and cooling system design. The procedures used are found in the American National Standards Institute (ANSI) vetted design manuals developed by the Air Conditioning Contractors of America (ACCA), and they are applied to Bob's single story "stick built" house.

Figure 1: Bob's house

Our fictional homeowner Bob, chose his contractor not because he was the least expensive, but because Bob felt comfortable this contractor would design, install and test the HVAC system using national building codes and the national standard for quality installations (ANSI/ACCA 5 QI – 2015 *HVAC Quality Installation Specification*, available at www.acca.org/quality). This would ensure his ENERGYSTAR™ rated system performed as it should.

MEET ACCA *Manual J*

Estimating how much heat a house will lose in the winter, or how much heat it will absorb in the summer, is based on the size and the type of building. If you have a big house, you will lose and gain a lot, and if you live where it's very cold or hot… you will gain or lose… a lot more.

However, to ensure that the right size heating and cooling unit will be selected, the first step that we'll discuss in the design process. Yes, the size, orientation and construction type of Bob's house matters, as

[2] HVAC – Heating, ventilation, and air conditioning.

does the place where he lives, which is Chicago, IL. But, as we survey the house, we will see that there are many other factors to consider.

A designer or HVAC contractor might get carried away, worrying about extraordinarily small minutia… or skip factors that have a very large impact on the home's heat loss/gain. That is why ACCA developed *Manual J*, to specify the items that must be considered, as well as the process, to determine how much heat is lost or gained in the house.

Bigger Is Not Always Better

A bigger pizza is better than a smaller pizza. Bigger diamonds are appreciated more than smaller diamonds. But bigger air conditioners and furnaces are not better, they are worse. Before we delve into what will be calculated, let's take a moment to look at WHY we're performing load calculations. So what if the heating and cooling equipment is too big? Does it really matter if the HVAC system is not properly sized? There are a number of problems that can be prevented by enforcing the minimum building design requirements as required by the building codes: excess expense, unnecessary wear, decreased comfort, and marginalized part-load performance.

First and foremost is MONEY. When heating and cooling systems are oversized, they require more money for installation. More money is needed for larger equipment, more money is needed for oversized copper refrigeration tubing or venting materials, more money is needed for larger ducts, and more money is needed for the labor to install all the extra stuff.

Second, oversized equipment starts and stops more than necessary. Heating and cooling systems are well made, but mechanical devices suffer the most trauma at start up.

- The cooling system is setting idle when suddenly a relay closes, a lot of power (about 240 volts & 40 amps) rushes into the compressor and fan motors in order to get them moving.
- A furnace is resting at about room temperature when suddenly burners ignite, and blue-hot flames broil the heat exchanger to about 180°F and expand the cooled metal.
- A duct system located in a hot attic or a cold crawlspace when suddenly warm or cold air comes rushing through it. The ductwork expands due to the air pressure and the joints and seams are stressed.

In addition, equipment starts are the highest single point of energy consumption. To illustrate, let me compare it to your car. Everyone knows their car will use more gas accelerating to 55 MPH than it will cruising at 55MPH for the same distance. Similarly, the startup of an HVAC system consumes a higher amount of energy. Once the HVAC system is cooling, or heating, the energy consumption drops below the peak value required to start the system.

As if increased cost and unnecessary wear on the equipment weren't enough, oversized furnaces and air conditioners often lead to an uncomfortable drafty home. Setting the thermostat at a lower or higher temperature may not make the home feel comfortable.

- Have you ever entered a home on a hot summer day and felt instantly cooler, but in just a few minutes you became less and less comfortable? Maybe you even felt warm again or "sticky".

The house may have smelled like a wet dog even when there was no dog present. This is another result of oversized equipment. When air conditioners operate for short periods of time, they drop the temperature but do not have time to remove the moisture in the air, so you feel clammy and uncomfortable.

- When air conditioners or furnaces run for short periods of time, the fan does not mix the air in the room and stagnant pockets of air form. The air also begins to stratify into different temperature levels, cooler air near the bottom and warmer air at the top. When that happens? People buy fans to stir the air or they turn the furnace fan on. Thus, more energy is used to try and get the basic comfort that should have been provided by a correctly DESIGNED HVAC system..
- If the heating or cooling system is centrally located in the house, then the warm or cool air is typically distributed evenly. However, when the equipment is located at one end of the house, that end will be cooled or heated first. When the equipment shuts off, the rooms furthest away will not be as comfortable.

High Humidity, Stagnant Air Pockets, Drafts, and Uneven Temperatures-Discomfort Times Four.

But wait, it gets worse… when a contractor performs a load calculation, the procedure is done in anticipation of hot or cold weather. What happens most of the time when it is warm, and not very hot outside? Or when it is only cool, but not cold? Every fall and spring when the temperatures are milder, those types of conditions occur. When they do, those problematic "oversizing" related issues will happen. Load calculations are performed to meet historical, 99%, worst case conditions based on temperatures recorded and averaged for a 30 year period. Thus, they are already "oversized" for mild weather; (say 85% load or 50% load conditions). Under milder conditions, oversized equipment short-cycles. Thus, most of the cooling and heating is done during the startup period when it is operating very inefficiently. As equipment oversizing is increased, short cycling related comfort problems become greater

A properly performed load calculation will ensure the heating and cooling equipment is sized based on a home's unique requirements. Surveying a home to gather the information to perform the heat loss/gain calculations can be time consuming, but it is time well spent when it can save money, promote prolonged equipment life, improve comfort, and conserve energy.

Why "Rules of Thumb" Don't Work

Proper heating and cooling equipment selection is certainly a prerequisite to achieving a comfortable home. Some HVAC contractors use an oversimplified "rule of thumb" to determine equipment size, which typically guesses to "a ton for every 500 or 600 square feet of floor area". Another common mistake for equipment replacement is to slap-in the same size of equipment by guessing the previous contractor sized it correctly. These selection guesses fail to take into consideration home building factors that greatly affect the choice of equipment and its capabilities, such as:

- *Windows:* How many? What size? What is the frame type? Does the glass have any special qualities like a Low-E rating? Which direction do the windows face? How much internal shading (blinds, drapes, etc…) do they have? How much protection does the external shading from overhangs, or porch roofs, provide for the windows?
- *Skylights:* What size? How many? What is the frame, glass and curb construction? Light shaft adjustment? What is the roof pitch where the skylights set? Which way do the skylights face (their orientation)?
- *Insulation:* Type and quantity in the walls, floors and ceilings (are all homes really the same?).
- *Duct System:* The location and design of the duct system can have a very large impact on the heating and cooling system size and efficiency.
- *Construction Materials:* Is the home constructed from light-weight "stick built" versus heavy-weight masonry walls?
- *Construction Practices:* How "tight" is the home? Were special products (house wrap) or procedures, (caulking, sealing) used to prevent excess energy usage?

These are just a few of the many factors that are critical to define and incorporate into the calculations used to determine the heating and cooling requirements for a home. Without taking all of them into consideration, equipment sizing is a guess. That guess may well cause problems for the homeowner, and in a worst case scenario, for the contractor too. More details are provided in Table 2 on page 7. Let's look at Bob's Floor plan:

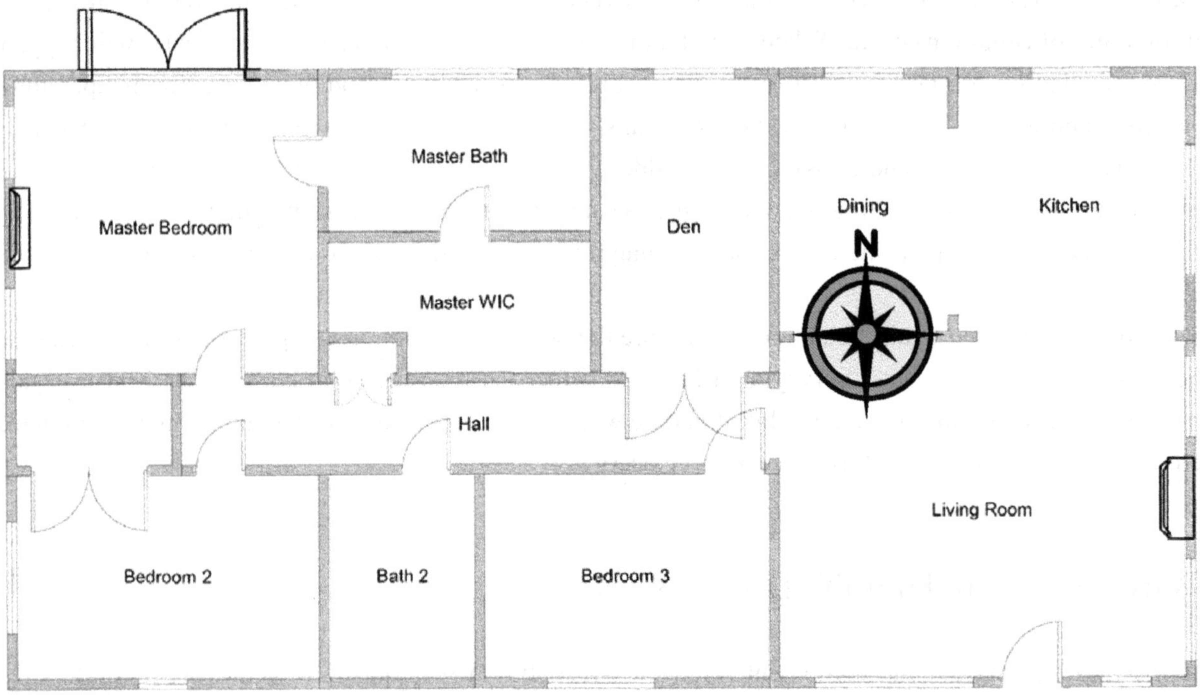

Figure 2: Bob's Floor plan

Contactors would normally perform heat loss/gain calculations for each room in Bob's house. But to simplify this analysis, we will concentrate on one area, Bob's master suite. This type of analysis is available when a room-by-room load calculation is done. Due to the different exposures to the outside, a room-by-room analysis is needed to provide the actual cooling and heating requirement for each room in the house. For simple equipment sizing where no duct replacement is anticipated, contractors will often take advantage of the opportunity to perform a "block load" calculation instead of the more detailed "room-by-room" analysis. Block loads get the same total value for equipment sizing. However, as the name implies, the block load calculation looks at the house as one big room, or "block", ignoring the individual room air flow rates. Note: Individual room-by-room load calculations provide room-specific air flow rates that are required in new construction and useful for diagnosing airflow problems in existing systems.

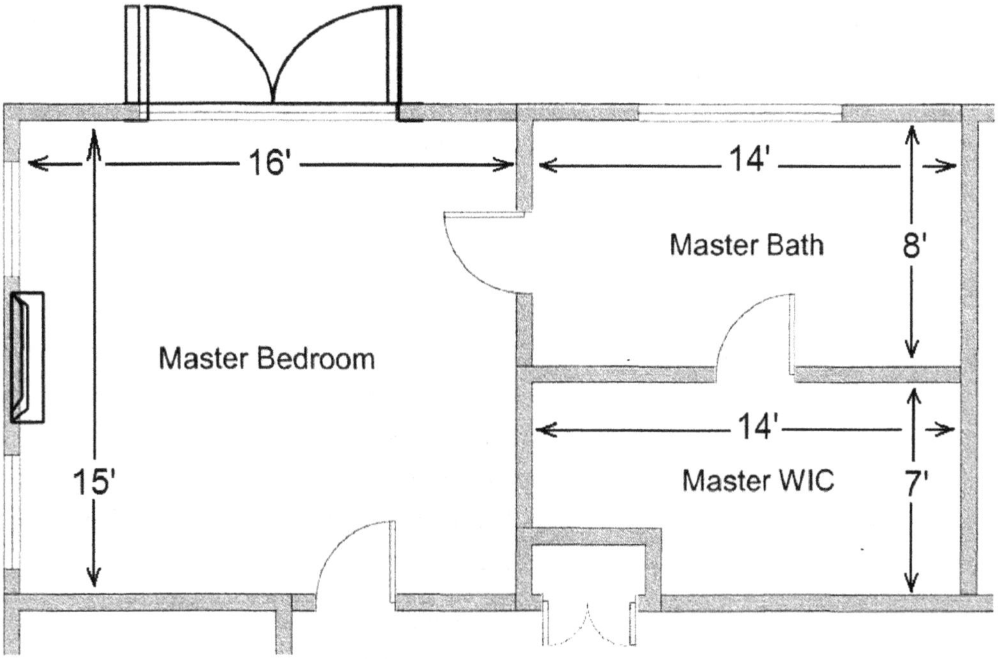

Figure 3: Bob's Master Suite

Design Information:

Annual weather conditions are different for various locations (i.e., Phoenix, Miami, Seattle, and Boston). One of the first things to consider is where Bob lives. *Manual J8* provides design weather data for almost every city and large town in the United States and Canada.

To begin, let's look at the design conditions for Bob's house: the home site is in Chicago Illinois. *Manual J8* offers design conditions for four different Chicago areas. The Chicago City Office area best represents where Bob's lot is, therefore the *Manual J8* design conditions stated in Table 1 below should be used:

Table 1: Weather Design Criteria

Season	Outside (MJ8 Table 1A)		Inside (MJ8 Defaults)	
Winter	2°F		70°F	
Summer	91°F	37 Grains†	75°F	50% Rh*

† For cooling, we use the specific moisture level of the outdoor air. Specific humidity is measured in grains of moisture. *Manual J8* discusses this at length.

* Later, when we select equipment, we need to know that 75°F and 50% relative humidity is equivalent to 63°F wet bulb temperature.

Manual J8 design conditions are based on decades of historical data and let us know that, on average, for only 1% of the year, the outdoor temperature is either hotter in summer or colder in winter than our design conditions. For indoor temperatures, settings where most occupants will be comfortable are based on the *Manual J* recommended 70°F for winter and 75°F for summer are selected. Design temperatures are target temperatures recommended by *Manual J.* Note: local codes may require the use of slightly higher or lower values. The selected HVAC system rarely matches the exact heating and cooling requirements, and there is usually extra capacity that allows for cooler or warmer temperatures. Standard design criteria are established because some homeowners may stay in a home for only a few years and then sell to someone else. Special indoor conditions provided for the original owner may not be suitable for next owners. For this reason, HVAC contractors need to carefully consider and document the justification for unusual design temperatures and/or humidity conditions.

Building Components Heat Gains and Losses

Heat is almost constantly passing through each component of the building thermal envelope in Bob's house (Figure 4). The heat transfer rate through these components (i.e., walls, windows, doors, floors etc.) influences Bob's heating and cooling requirements, also known as the house's load. During the heating season, the HVAC system must add heat to Bob's house to maintain the desired temperature, and during the cooling season, heat must be removed.[3] The amount of heat that transfers depends on the Heat Transfer Multiplier (HTM) and the area of the building component (Load = HTM × area).

The HTM is determined by the building component's (windows, walls, ceilings, etc.) heat transfer traits and the temperature difference on each side of the building component. *Manual J8*'s tables provide HTM values for different building components. For example, the heating HTM for the wall in Bob's house equals the heat conduction value (U value) of the wall components, multiplied by the design temperature difference (TD). Bob's wall HTM = 0.086 (Table 4A - sheetrock, insulation, framing, etc.,), multiplied by the winter TD (TD = 70°F indoors - 2°F outdoors = 68°F). Therefore, this particular wall assembly's heating HTM = 0.086 x 68°F = 5.85.

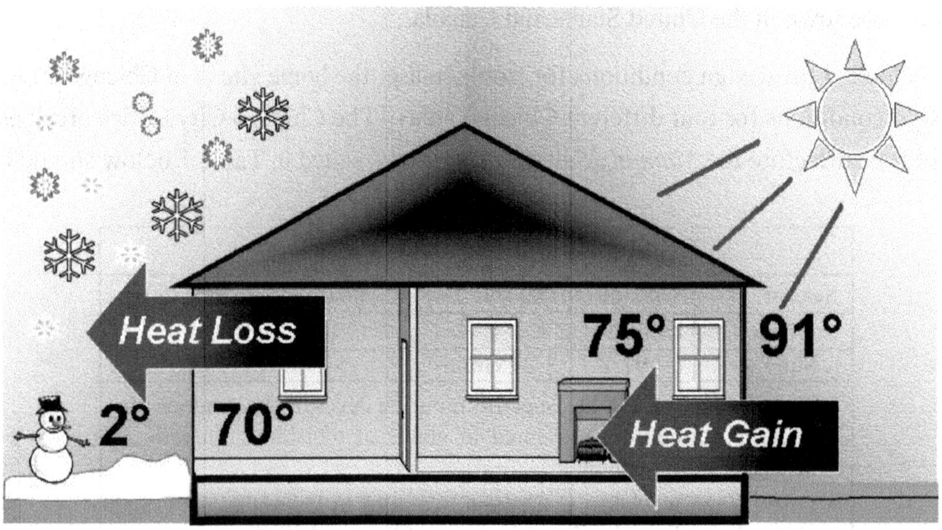

Figure 4: Heat Transfer In and Out of a Home

[3] Technically, HVAC systems do not "cool", they remove heat.

The heating and cooling loads contributed by each building component of Bob's master suite (bedroom, bathroom, and walk-in closet), using the *Manual J8* process, are outlined in Table 2. Here you can see that some building components have very high heat transfer multipliers.

Table 2 illustrates the building components, their description, HTMs, area, and loads. It identifies load sources and shows their relative contribution to total heat loss / heat gain. (*Manual J8* provides detailed instructions, examples, tables, and forms. Software for the abridged version of *Manual J8* is available at www.acca.org/speedsheet).

Table 2: Master Suite Portion of the Component Loads

Part of the Home	Description	Heat Transfer Multiplier (HTM)*		Area	Winter Heat Loss (in Btu/h)‡	Summer † Heat Gain (in Btu/h)‡
		Heating	Cooling			
Windows	Bedroom: two - 2'6"x 4,' facing West Bathroom: one - 6'6" x 1' facing North Windows: wood frame, double pane, low-E, (Drapes: light color, ½ closed) , no overhang	49.02	59.55	26.5 sqft	**1,299**	**1,578**
		(Windows – 3a-2ow, MJ8 Table 3, and Worksheet B)				
French door	8' French door: wood, double paned, low-E, facing North and has light colored blinds, usually half closed, 6" overhang	41.18	12.29	56 sqft	**2,306**	**688**
		(French Door – 10C-w, MJ8 Table 3, and Worksheet B)				
Ducts	R8 insulation, trunk and branch, located in an open crawl space; the ducts are only partially sealed	This load a function of envelope load and is calculated on Worksheet G		Default duct surface area used	**2,446**	**978**
		(Ducts – 7G-R, MJ8 Table 7, and Worksheet G)				
Ceiling	Under vented attic (with R30 insulation) roof has dark asphalt shingles	2.18	1.61	450 sqft	981	726
		(Ceiling - 16B-30ad, MJ8Table 4A and Worksheet D)				
Floor	Over an open crawl space (with R19 insulation in floor joists)	3.40	0.55	450 sqft	1,530	250
		(Floor - 20P-19w, MJ8 Table 4A and Worksheet D)				
Walls	Only walls exposed to the outdoors (with R15 insulation)	5.85	1.22	284 sqft	1,661	348
		(Walls - 12D-0sw, MJ8Table 4A and Worksheet D)				
Internal gains	Bob and his wife (at rest) Electronic equipment - computer and fax/copier/scanner.	See Worksheet F and the equipment manufacturer's product data			0	**1,060**
		(Internal Gains, MJ8 Table 6a & b and Worksheet F)				
Other	Infiltration and other items	From MJ8 Tables and Worksheets			**3,909**	509
		(MJ8 Tables and Worksheets)				
				Total	14,132	6,137

* HTMs are based on the building components heat transfer characteristics and the temperature difference.
‡ Appropriate HTM × area = Winter heat loss and Summer heat gain.
† Summer heat gain is concerned with Sensible load only. sensible and latent loads are discussed on page 13

So What Does All That Mean?

This winter, Bob's house will lose a significant amount of heat due to the cold winters in Chicago (see Figure 5 next page). During the heating season, when the temperature difference between the inside, and outside of the home is 68°F, the components allowing the greatest heat loss are:

- *Infiltration* – This is uncontrolled air that leaks through the cracks and holes in the building's envelope, 28% of the total load. Infiltration can be induced by external forces such as wind forcing the outside air into the home, or negative pressures within the home induced by bathroom exhaust fans, clothes dryers, and kitchen exhausts pulling air into the home to replace what was exhausted. Holes can be as obvious as an open fireplace damper, or as inconspicuous as the gap at the strike plate of a door jamb. Weatherproofing or sealing those air paths could save a significant portion of the infiltration heat loss in Bob's master suite.

- *Glass* – Windows and French door – Glass can allow massive amounts of heat to be transferred outdoors; 26% of the total heat loss in Bob's master suite is through glass. The heat loss through the master suite windows amounts to 9% of the total loss, and the French door accounts for an additional 17%.
- *Ducts* – This heat is lost to the crawlspace. Duct loss should be minimized by thoroughly sealing and insulating. However, if the ducts were relocated from the crawl space to the conditioned space, 17% of the heating load would be eliminated.

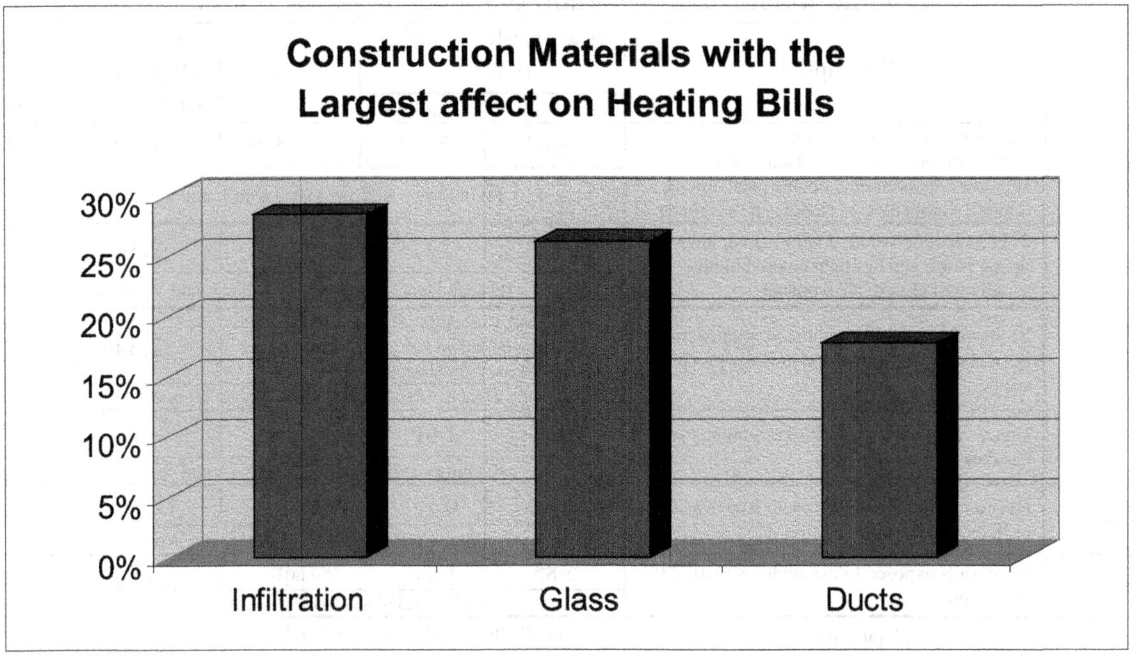

Figure 5: Heating Bill Factors

During the cooling season (see Figure 6 next page), the components allowing the greatest heat gains are:

- *Glass* – Windows and the French door – This is another big load, 37% of the cooling capacity. The heat gain through Bob's windows is more than a third of the master suite's cooling load (and cooling cost).
- *Internal Loads* – Bob, his wife, and the office equipment/electronics create internal gains in the master suite. These gains should not be overlooked because they can make a significant contribution to the total load, in this case 17%.
- *Ducts* – In the summer, ducts absorb heat from the open crawl space, about 16% of the total heat gain. If the ducts are poorly insulated they can absorb even more heat and excess moisture[4]. This added load will cause the cooling equipment to run longer because the ducts deliver less conditioned air to the room.

[4] Heat and moisture added to return air increases the sensible and latent load on the cooling coil and affects the wet bulb temperature of the air entering the coil. This affects performance of the cooling coil, and reduces capacity available to cool the space.

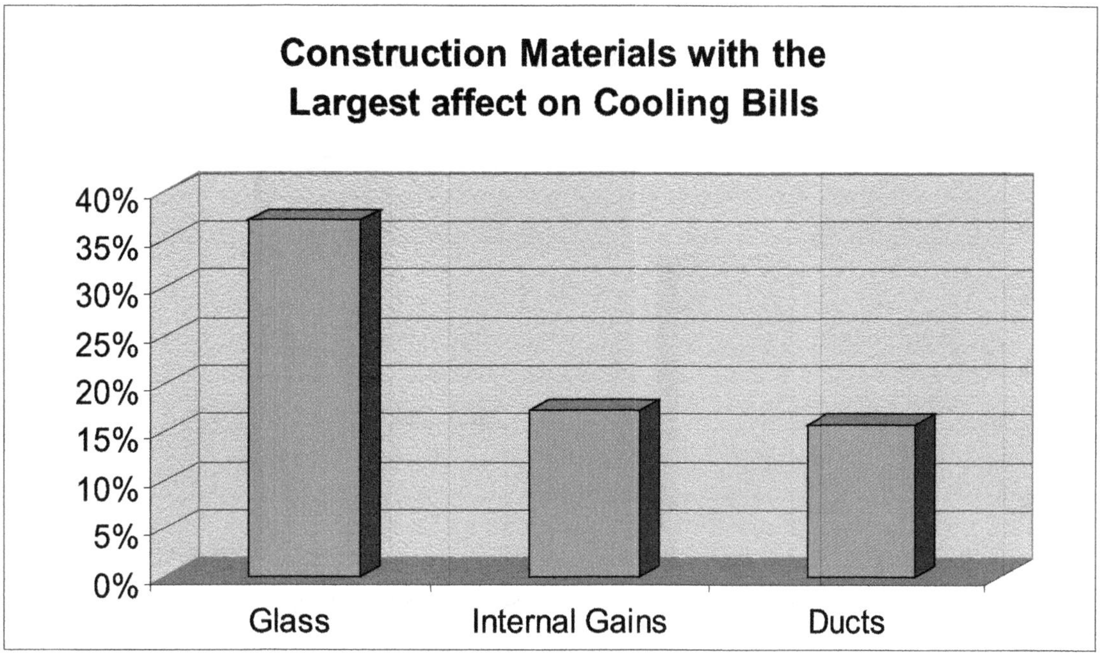

Figure 6: Cooling Bill Factors

Table 3 shows the results from completing the calculations for the remaining portion of Bob's house and provides a heating and cooling load comparison of the size of the space to the amount of the load.

Table 3: Load Comparison – House vs. Master Suite				
Load	Bob's House	Bob's Master Suite	Percentage Size	Percentage Load
Heating	51,838 Btu/h	14,132 Btu/h	25%	27.3%
Cooling †	21,423 Btu/h	6,137 Btu/h	25%	28.6%

† This is the sensible heat gain, the main concern at this point. A discussion of the different types of heat, sensible and latent, is discussed on page 16

As Figure 7 (next page) illustrates, square feet of living space (area) is a poor indicator of the heating and cooling capacity requirements. Bob's master suite occupies 25% of the home; however, its heating and cooling requirements are about 27% and 28%, respectively, of the whole house's conditioning needs. You can see that HVAC contractors cannot simply, or correctly, guess the load and then proportion the equipment capacity based simply on the size, or area, of a room.

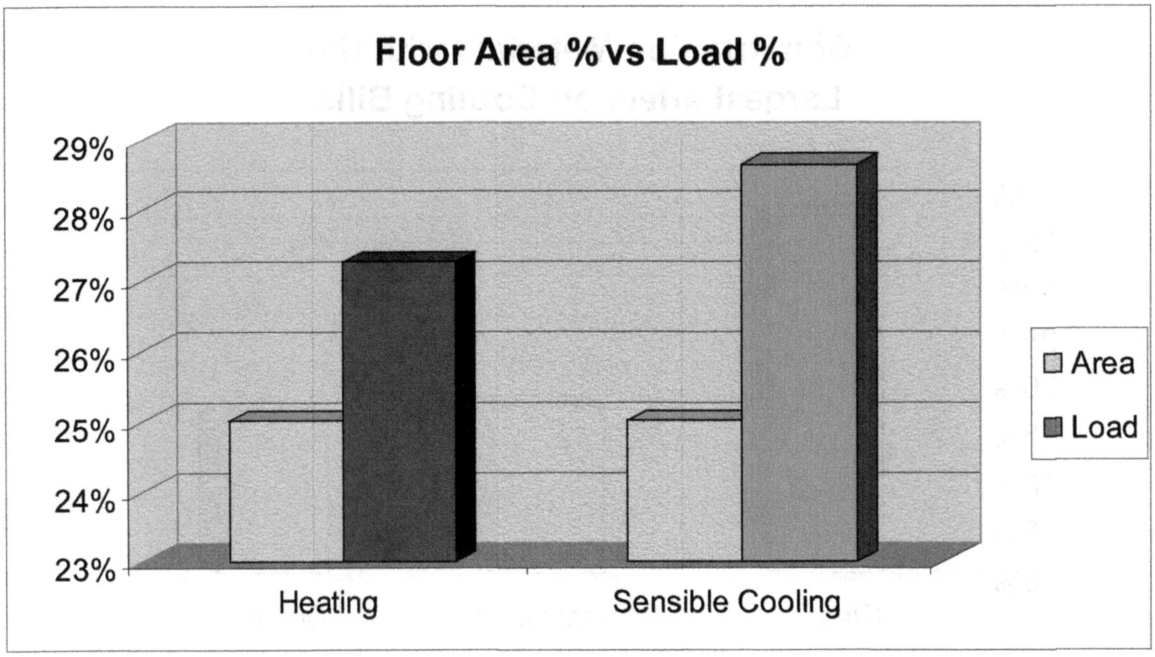

Figure 7: Bedroom Floor Area Compared to the Bedroom Load

We have now completed the first step of the design process. Table 4 summarizes the results produced by our load calculation using *Manual J8*. This information and HVAC equipment manufacturer's performance data will be used to select equipment.

Table 4: Load Calculation – Design Data Summary			
Season	Load	Outdoor Conditions	Indoor Conditions
Heating	51,838 Btu/h	2°F	70°F
Cooling†	21,423 Btu/h	91°F	75°F; 50% RH
† The cooling load is primarily concerned with sensible heat only at this point. A discussion of the different types of heat, sensible and latent, is discussed on page 16			

LOAD CALCULATION SUMMARY

An HVAC contractor must use the correct set of design conditions for a home to calculate the heating and cooling loads. Building materials and construction practices impact a home's heating and cooling needs. However, some have a greater overall effect on the total load than others. Each region of the country will have a set of building components that need more attention than others because they account for most of the load (see Appendix A). When reviewing a permit application (code official) or contractor's proposal (homeowner), consider the following design factors. These have been weighted to reflect their impact on the heat loss and heat gain calculations.

Load Calculation Checklist

Impact	Issue
High	Ensure a rigorous load procedure (MJ8) was used to calculate requirements
	Design Temperatures (Verify Table 1A for local outdoor standard; MJ8 indoor default or local indoor standard)
	Windows, Glass Doors and Large Skylights (Examine number, size, exposure directions, U-factor and SHGC, external/internal shade)
	Ducts (Examine location, leakage and duct wall R-values)
	Ceilings under an attic (Ensure R-values, roof material, roof color)
Medium	Infiltration (Verify stated tightness for load calculation matches observed tightness)
	Ventilation† (Look for heat or energy recovery equipment or ducts carrying outdoor air connected to the return plenum)
	Small Skylights
Low	Appropriately Insulated Floors (Verify R-values, floor material, and under floor type, e.g. slab)
	Appropriately Insulated Walls (Verify R-values, wall construction material)
	Internal Gains (normally low, about one person per bedroom and two people in the master bedroom, but some people have hobbies or electronic needs that make this a big cooling-load item)

† Infiltration is not engineered ventilation. If code requires ventilation, there shall be a ventilation load separate from the infiltration load.

Figure 8: Load Calculation Checklist

Manual J8 load calculations should be aggressive, which means that the designer should take full advantage of legitimate opportunities to minimize the size of the estimated loads. The practice of manipulating the outdoor design temperature, failing to take full credit for efficient construction features, or ignoring window shading devices, and then applying a "safety factor" will lead to inaccuracies later in the process. Research studies indicate the *Manual J8* procedure provides an adequate safety factor. No additional safety factors are required when load estimates are based on accurate information pertaining to envelope construction and duct system efficiency. Further, HVAC systems rarely meet the exact heating and cooling requirements, so there is usually extra capacity that allows for cooler or warmer temperatures.

CHAPTER II
MANUAL S: SELECTING HVAC EQUIPMENT FOR COMFORT

Do You Need a Sledgehammer to Swat a Fly?

The next step is to find the right hammer to strike Bob's heating and cooling loads. We need to choose HVAC equipment that will maintain the design comfort conditions we discussed previously. We must ensure that the HVAC equipment is compatible with this dwelling's heating and cooling needs.

We may select heating equipment using electric, gas, oil or propane. Typically, cooling equipment is electric only. Geographic location and available energy sources may sometimes direct your equipment choices. There are many combinations:

For Heating:

Forced Air

- Fossil fuel furnaces (Gas or Oil)
- Stoves burning wood or fuel pellets
- Air source heat pump
- Electric furnace or resistance heat
- Geothermal or water source heat pump

Hydronic Heat

- Boiler (hot water or steam heat)
- Geothermal or water source heat pump
- Electric water heater
- Fossil fuel water heater

For Cooling:

Forced Air

- Electric Air Conditioner
- Air source heat pump
- Evaporative Cooler (Swamp cooler)
- Geothermal or water source heat pump

Give 'em the Heater Charlie

Most of these different systems could be used in almost any combination. Bob selected a gas furnace and an electric air conditioner. From our load calculations, let's review Bob's heat loss information and target **heating** capacity requirements:

Table 5: Heating Design Information		
Mode	Load	Design Outdoor Temp
Heating	**51,838 Btu/h**	2°F

The furnace we need must produce 51,838 Btu/h of heat. If we look at the manufacturer's product data in Figure 9 (next page), we see that a furnace like the XYZ model **FR60-036** has an input capacity of 60,000

Btu/h, but its output capacity is only 48,000 Btu/h. If we install this "60,000 Btu/h" furnace in Bob's home, his family will be cold when the outdoor temperature drops to the design temperature of 2°F.

Input Capacity | Output Capacity

XYZ Furnace Company						
General Data						
			Efficiency			
Unit Size	FR60-024	FR 60-036	FR 75-024	FR 75-036	FR 75-048	FR 75-060
Output Capacity:						
Upflow	48,000	48,000	60,000	60,000	60,000	60,000
Horizontal	48,000	48,000	60,000	60,000	60,000	60,000
Input Btu/h	60,000	60,000	75,000	75,000	75,000	75,000
Temp Rise Range	30 – 60	15 – 45	45 – 75	30 – 60	40 – 70	30 – 60

Figure 9: XYZ Model FR60-036 (60,000 Btu/h Input and blower Capacity for 3 ton AC)

The ABC model **FU60-36** furnace, Figure 10, has an input of 60,000 Btu/h, but its output is 56,000 Btu/h. The furnace has a bit too much capacity, about 8% (56,000 ÷ 51,838 = 108%), but this is acceptable[5].

Output Capacity | Input Capacity

ABC Furnace Company						
Performance Data						
			Efficiency			
Unit Size	FU60-024	FU60-036	FU80-0[illegible]	[illegible]	[illegible]U80-048	FU80-060
Output Capacity:						
Upflow	56,000	56,000	56,000	[illegible]	[illegible]0,000	70,000
Horizontal	56,000	56,000	56,000	56,000	70,000	70,000
Input Btu/h	60,000	60,000	60,000	60,000	80,000	80,000
Temp Rise Range	30 – 60	35 – 65	45 – 75	30 – 60	40 – 70	30 – 60

Figure 10: ABC Model FU60-36 (60,000 Btu/h Input and blower Capacity for 3 ton AC)

This ABC model FU60-036 furnace has an OEM design temperature rise range of 35°F - 65°F. The air temperature rise through the furnace depends on the airflow rate[6] (Cubic Feet per Minute – CFM) through the heat exchanger. If the airflow is outside the temperature rise range, then: the equipment may cycle off on safety limits, the furnace could possibly suffer damage, or possibly even create an unsafe condition. Here are some problems that can occur when the airflow is incorrect:

- Slow Air (too much temperature rise): Slow moving air may allow the heat exchanger to become too hot, which may damage (warp or crack) the metal heat exchanger.
- Fast Air (too little temperature rise): Excessive airflow rate; air moving too quickly may cause condensation in the metal heat exchanger (this produces an acid that can harm or penetrate the heat exchanger.).

---

[5] For furnaces, ACCA *Manual S* says output capacity may exceed the *Manual J* heating load by 40%.

[6] Air volume is measured in CFM. CFM is the amount of air in cubes (one foot by one foot by one foot) which move through a duct in one minute.

The ABC FU60-036 will meet Bob's heating needs, so now we need to look at its airflow. Below we see another extremely useful and important table, the fan performance table for the ABC FU60-036. Different size duct systems restrict airflow differently. You can blow more air through one foot of garden hose than through a soda straw. Likewise, the furnace fan can blow more air against low resistance than through high resistance. This table will demonstrate how much air the unit can move after factoring in all forms of airflow resistance (filters, duct system, etc.). Low resistance, like the garden hose, is about 0.2 inches of water column[7] (IWC). If the blower meets higher resistance (for example, 0.7 IWC [soda straw]), it moves less air.

If we were selecting only heating equipment, we would chose a furnace which delivers the target CFM against a low total system resistance, 0.30 IWC or less. This assumes the duct runs are sized properly and there are no extra filters or other accessories that produce significant resistance to airflow. When selecting a furnace, or fan coil, we want to ensure the blower can deliver its target airflow when working against a total system resistance of 0.50 IWC or more. When making the initial equipment search we look for equipment that can deliver the desired airflow at a ***medium*** level fan speed.

ABC Furnace Company FU60-036								
Air Delivery – CFM (With Filter)								
Unit Size	Speed	External Static Pressure (inches water column)						
		0.1	0.2	0.3	0.4	0.5	0.6	0.7
FU60 - 024	High	1075	1040	995	945	895	840	760
	Med – Hi	950	925	895	845	795	740	660
	Med – Lo	850	825	780	740	685	635	560
	Low	740	700	650	620	565	515	455
FU60 - 036	High	1470	1415	1400	1285	1[illegible]15	1120	995
	Med – Hi	1315	1280	1235	1298	1[illegible]15	1035	930
	Med – Lo	1125	1110	1085	1045	995	915	830
	Low	930	925	910	850	830	770	705
FU60 - 048	High	1700	1685	1640	1580	1545	1450	1380
	Med – Hi	1500	1465	1435	1385	1255	1300	1250
	Med – Lo	1325	1295	1265	1230	1190	1150	1105
	Low	1205	1170	1145	1110	1080	1035	990
Note: Airflow based on a dry coil at 120v with factory approved filter.								

Figure 11: OEM Blower Performance Data

From the manufacturer's blower performance table, Figure 11, we see the FU60-036 unit can deliver 995 CFM on medium-low speed when working against 0.5 IWC of total system resistance. It remains to be seen if this is adequate for the cooling equipment. If we discover the blower performance is deficient at any stage in the design procedure, we must go back and select a furnace with more blower power (or accessories with less resistance). This also applies to blowers installed for cooling and heat pump applications. Keep in mind:

[7] The measure, inches of water column (IWC), comes from the amount of pressure used to force water up a column.

- We will need to size the duct runs (in the next step using ACCA *Manual D*) to know if our assumption about blower performance is correct.
- By selecting a mid-range airflow, if the actual installation differs slightly from the design, the installer still has an opportunity to adjust the airflow by increasing or decreasing fan speed.

For heating, 995 CFM is acceptable if it ***falls within the temperature rise range the manufacturer prescribed***. Our heating airflow is 995 CFM, which equates to a furnace temperature rise of 51°F:

$$\Delta T = Btu/h \div (CFM \times 1.1 \times ACF)$$

$$51°F = 56{,}000 \div (995 \times 1.1 \times 1)$$

Where:

ΔT = Temperature difference in the air between the inlet and outlet of the furnace or cooling coil.
Btu/h[8] = Thermal output capacity of the furnace or cooling system.
CFM = Cubic feet per minute; volume of air moved through the furnace by the blower assembly.
1.1 = A rounded off physical air constant (derived from laws of physics), adjusts the units properly (Note: 1.08 is often used for this calculation as well).
ACF = Altitude Correction Factor; an adjustment for air density at differing altitudes (1 applies to sea level; for example, at 2000 ft the ACF is 0.93 and at 10,000 ft. ACF is 0.69).

51°F is very close to the mid-point operating temperature range specified by the original equipment manufacturer's (OEM) – 50°F[9]. The ABC FU60-036 will meet our heating needs (for heating capacity and temperature rise).

ΔT @ 51°F is a useful number for establishing the correct airflow across the coil using the OEM table once the equipment is installed. Additionally, once the temperature increase in the discharge air is known, a simple engineering formula can be used to double check the sizing of the equipment.

$$Btu/h = (1.1 \times CFM \times ACF \times \Delta T)$$

$$55{,}820 = (1.1 \times 995 \times 1 \times 51) \text{ and } 55{,}820 \div 51{,}838 \times 100 = 108\% \text{ of design}$$

Where: (same as above for ΔT, Btu/h, CFM, 1.1, and ACF)

Note: If Bob had selected a fossil fuel furnace with a multi-stage or modulating burner, the sizing would not change. Equipment sizing requirements in ACCA Manual S are based on the maximum output of the furnace. One common misconception is that by modulating the burners, or staging the cooling compression levels, the furnace or AC control system will "self-correct" for sizing errors. The truth is, if the ducting and other air distribution components need to be upsized to accommodate the larger piece of equipment, the system will be difficult to balance (if not impossible), and will never operate at the design efficiency the homeowner paid a premium price for.

Now it is time to select the cooling equipment.

---

[8] In this case, we use the furnace's actual capacity in the formula instead of the *Manual J* heat loss. The actual furnace output capacity of 56,000 Btu/h is used, not the 51,838 Btu/h design capacity from the load calculation.

[9] CAUTION: Furnace manufacturers specify a minimum and maximum acceptable temperature rise for each make and model. Operating equipment at flow rates lower than performance data table values may cause problems. 50° is the mid point of 35° - 65°; see the temperature rise performance data in Figure 10.

And Now for Something Completely Different

Bob's home will also have air conditioning, so let's review his cooling target information. Table 6 summarizes the requirements the air conditioner needs to be able to deliver 26,107 Btu/h of total cooling: 21,423 Btu/h of this load is sensible and 4,684 Btu/h of the load is latent.

Table 6: Cooling Design Information					
Mode	**Total Load**	Sensible Load	Latent Load	Design Outdoor Temp	Design Outdoor Humidity
Cooling	**26,107 Btu/h**	21,423 Btu/h	4,684 Btu/h	91°F	37 Grains

Sensible and Latent Loads

When considering cooling load, there are two parts - the sensible and latent load.

The sensible load is the heat that is measured by a thermometer or a thermostat. It is heat that you consciously feel. It is the heat you associate with a desert or a dry heat.

Latent load is the heat associated with airborne moisture (produces relative humidity as measured by a hygrometer or humidistat). During the cooling process, moisture is extracted from indoor air.
High latent heat or high relative humidity feels humid, clammy, or steamy. An example is a home that shows a cool temperature on the thermostat but, because of high humidity, still feels uncomfortable. When you walk into this home your initial impression is that it is cool, but then as your body adjusts to the room temperature, you begin to feel sticky, clammy, and uncomfortable. You may even feel warm again. This is why two homes that have the same reading on a thermometer or thermostat can feel very different.

The *Manual J8* procedure provides the home's cooling requirements in two forms of heat: sensible and latent. Together they add up to the total heat gain or total cooling load. To help select the right HVAC unit, it is helpful to know the ratio of sensible heat to total heat in the home (sensible load divided by the total load). This ratio is called the home's sensible heat ratio (SHR).

The home's SHR informs us of the requirement for moisture removal. Bob may need a unit with high sensible and low latent cooling (more temperature control and less moisture removal). He may need his new unit to concentrate on removing moisture as much as controlling the temperature. The home's SHR is a valuable design tool, but many times is overlooked in the design process by poorly trained HVAC contractors. In the ACCA procedures, it is used to select a target value (preliminary, ball-park figure) for airflow moving through the air conditioner's cooling coil. This airflow rate helps to narrow down the choices as the selection process begins [10].

- Low airflow rate – (better suited for reducing high indoor relative humidity)
    - This allows the indoor refrigeration coil to get very cool, which allows more moisture to be removed from the air,
    - Low airflow is appropriate for humid climates,

[10] In general, as airflow increases, an air conditioners sensible capacity increases and its latent capacity decreases, and vice versa (i.e., more airflow, higher indoor humidity; less airflow, lower indoor humidity).

   - Low airflow is equivalent to 350 to 400 CFM per ton of total cooling capacity,
   - For example, the airflow rate for a 2.5 ton unit could be 875CFM[11]
     2.5 (tons) × 350 (CFM) = 875CFM.
- Medium airflow – (best suited for homes with average indoor relative humidity)
   - This flow rate meets the indoor temperature and moisture removal needs for most of the country,
   - Medium airflow generally varies between 400 and 450 CFM per ton of total cooling capacity,
   - For example, the airflow rate for a 2.5 ton unit could be 1,025 CFM
     2.5 (tons) × 400 (CFM) = 1,000 CFM.
- High airflow [12] – (better suited for homes with little or no indoor relative humidity)
   - This airflow does the mostly "sensible" cooling (up to 100% sensible cooling). The air moves quickly, cooling the home, but it not removing much moisture.
   - High airflow is very useful in dry climates, where the latent load is very small or non-existent.
   - High airflow is generally considered at or above 450 CFM per ton of total cooling capacity.
   - For example, the airflow rate for a 2.5 ton unit could be 1,125CFM
     2.5 (tons) × 450 (CFM) = 1,125CFM.

400 CFM per Ton Myth

The 400 CFM per ton rule of thumb that is so popular with HVAC contractors has nothing to do with Bob's house (or any other home, for that matter). It generally keeps refrigerant temperatures and pressures in an acceptable range, or it may be the value used to rate the equipment in a laboratory test chamber, but that has nothing to do with Bob's family's comfort. Proof of the fallacy of this 400 CFM per ton 'magic' is provided by equipment manufacturer's data for indoor-outdoor combinations that operate at less than 400 CFM per ton to more than 600 CFM per ton.

If the home's SHR is high (0.85 or more), it will be relatively easy to select cooling equipment that will control indoor humidity. If the home's SHR is medium (0. 80 to 0. 85), then there is more humidity (latent load) to deal with, and it becomes more important to select cooling equipment that will control indoor humidity. If the home's SHR is low (0.80 or less), there is an unusually large latent load and it may not be possible to find (standard) cooling equipment that will control indoor humidity. Supplemental dehumidification may be necessary.

As already noted, the *Manual J8* SHR is used to determine the target airflow. The SHR for Bob's house is 0.82:

---

[11] CAUTION: Cooling equipment manufacturers specify acceptable flow rates for each make and model. Operating equipment at a lower flow rate may cause coil icing, undesirable refrigerant pressures, and possible equipment damage. Operating equipment at higher flow rates can cause condensate to blow out of the drain pan.

[12] At the high end, airflow rate is limited by the possibility of condensate blow-off. If the air velocity through a wet cooling coil is too high, water will be blown into the supply plenum or duct. If this happens, there is a high risk of mold and mildew in the supply duct. (Desert climates do have a rainy season where the cooling coil may be wet for a week or two.) Refer to equipment manufacturer's performance data for the maximum airflow velocity through the cooling coil.

Sensible Heat Ratio = Sensible Load ÷ Total Load

0.82 = 21,423 Btu/h ÷ 26,107 Btu/h

Table 7: Bob's Home's SHR		
Load	**Bob's House**	**Percentage of Load**
Sensible (Heat)	21,423 Btu/h	82%
Latent (Humidity)	4,684 Btu/h	18%
Total	26,107 Btu/h	100%

When we have the SHR for Bob's house, using this ACCA *Manual S* table provides a preliminary value for cooling CFM.

Table 8: Sensible Heat Ratio vs. Cooling Coil Temperature Difference (ΔT)	
Home's SHR	ΔT
Below 0.80	21°F
0.80 – 0.85	19°F
Above 0.85	17°F
Temperature Difference = Entering dry bulb – Leaving dry bulb	

Bob's home's SHR was 0.82, which equates to a 19°F temperature difference across the cooling coil. By using this formula, we can calculate a *target* airflow or fan speed for Bob's cooling system.

CFM = Sensible Load ÷ (1.1 × ACF × ΔT)

1,025 CFM = 21,423 Btu/h ÷ (1.1 × 1.0 × 19)

Where:

Sensible Load (Btu/h) = The sensible cooling load from the *Manual J8* load calculation.

ΔT = Temperature difference in the air between the inlet and outlet of the furnace or cooling coil.

Btu/h = Thermal output capacity of the furnace or cooling system.

CFM – Cubic feet per minute; the volume of air moved by the blower assembly in the furnace.

1.1 = A physical air constant (derived from laws of physics: rounded up from 1.08 for ease of use), adjusts the units properly.

ACF = Altitude Correction Factor, an adjustment for air density at differing altitudes (1 applies to sea level). Bob's house is less than 1,000 feet above sea level, so we will round it to 1.0.

So for Bob's house, our cooling airflow ***target*** is **1,025 CFM.** Commercial designers can select custom equipment especially for their design criteria. The residential HVAC contractor does not have as much flexibility. Equipment is manufactured to operate at one or two fan speeds[13]. We have to find the closest match in the manufacturer's product line. This 1,025 CFM value is just the target airflow rate we use in the cooling equipment performance data. After we select our equipment, we may find that its airflow rate is slightly different (± 15%).

[13] Fans may have three or more speed settings, but the unit operates at one speed for cooling and possibly another for heating. For furnaces, the cooling fan speed is usually higher than the heating fan speed. For simplicity, Bob's house will use one fan speed in heating and cooling; follow the installation instructions for your unit.

Small, Medium, or Large?

There are two main sources for air conditioner and heat pump performance data: The Air conditioning, Heating, and Refrigeration Institute (AHRI) ratings, and the original equipment manufacturer (OEM) performance data.

AHRI tests and publishes information pertaining to the "data plate" seasonal efficiency of electric air conditioners and heat pumps. The AHRI published efficiency and capacity ratings are based on the following special set conditions:

- The unit is tested at 400 CFM per ton,
- The outdoor dry bulb (DB) temperature is 95°F,
- The DB temperature of the air entering the cooling coil is 80°F,
- The indoor/entering wet bulb[14] (EWB) temperature is 67°F (equivalent to 80°F DB and 50% Rh).

AHRI ratings are a great way to ensure standardization, to compare the performance of different pieces of equipment, and to "zero-in" on the approximate equipment size. However, the AHRI rating guide is an inappropriate resource for selecting cooling equipment. Here are some reasons why:

- Many parts of America are very dry, or very humid, most of the time, so the correct airflow rate for a particular home may range from 350 CFM per ton to 450 CFM or more per ton. Remember, cooling equipment capacity depends on the airflow rate through the cooling coil.
- Many American cities and towns may actually have an outdoor design temperature of 95°F, but most have a lower or higher outdoor design temperature (*Manual J8* Table 1 depicts temperatures from 80 °F to 115°F).
- The last item is the AHRI value for entering[15] or "indoor" temperature. Very few people like an 80°F living space, even if the relative humidity is 50%.

This is why it is extremely important when selecting cooling equipment to use the OEM application (or performance) data instead of the AHRI rating guide. For example on the next page, Table 9 and Figure 12 compare AHRI test conditions with the real design conditions in different cities.

Table 9: AHRI Data (Standard 210/240) Comparison to *Manual J8* Weather Data

	AHRI	Chicago	Phoenix (Arid)	Miami (Humid)
Airflow/Ton	400 CFM	400 CFM	450 CFM (High SHR)	350 CFM (Low SHR)
Outdoor Drybulb	95°F	91°F (-4°F)	106°F (+11°F)	89°F (-6°F)
Entering Wet Bulb	67°F (80°F/50%)	63°F (75°F/50%) Lower EWB	57°F (75°F/30%) Lower EWB	65°F (75°F/70%) Lower EWB

AHRI does not consider grains of moisture in their evaluation criteria. However, the grains of moisture present during the design conditions will affect the entering wet bulb (EWB) by infiltration, and therefore

14 75°F and 50%Rh are equivalent to 63°F Entering Wet Bulb. Wet bulb temperature takes into account both the temperature measured on a thermometer and the moisture content of the air.

15 The condition of the air entering the cooling coil depends on the condition of air returning from the space. If the return ducts are leaky, poorly insulated, and are in a vented attic or open crawl space, the entering air will probably exceed 80 °F dry bulb and 63 °F wet bulb. This is very important, because equipment capacity is very sensitive to the condition of the entering air.

the latent capacity of the equipment. The charts (in Figure 12) are a graphic representation of the data on Table 9. The AHRI data, as seen here, does not represent a number of cities. Consult Table 1A or 1B in *Manual J8* to find the actual operating conditions in your city or area.

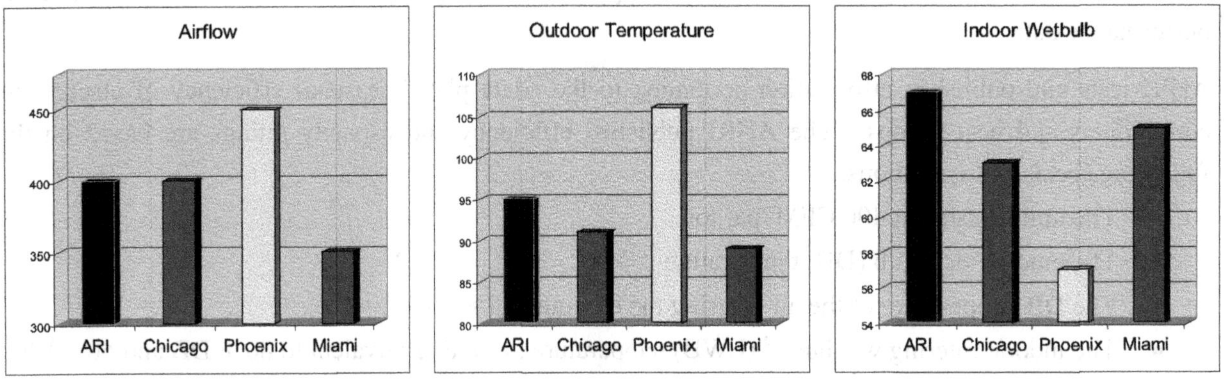

Figure 12: Charts Depicting AHRI Data Compared to Three Cities

Because different people prefer different temperatures and humidity levels, it can be hard to select design conditions. This is made more difficult because people rarely stay in one home all of their adult lives. The Smith family may build the house and sell it to the Johnson family in a few years. The Smiths and Johnsons may enjoy different temperature and humidity levels. This is why it is best to select one standard set of design conditions most likely to be agreeable with the general public. ACCA *Manual J8* recommends[16]:

- Using the design outdoor temperature from *Manual J8*, Table 1A (or 1B) for your city,
- Calculating the design airflow,
- Using 75°F and 50% Relative humidity for summer indoor design conditions.

The OEMs produce detailed performance data for their cooling equipment (air conditioners), like they do for their heating equipment (furnaces). You generally need to know three design factors when selecting an air conditioner (or heat pump in the cooling mode):

- <u>Cooling coil airflow</u> – Generally expressed in these blower speed intervals:
  - Low - High,
  - Low – Medium – High,
  - Low – Medium Low - Medium High - High.
- <u>Outdoor air temperature</u> – This is the temperature of the air around the outdoor condensing coil[17], sometimes called the condenser air temperature. (Keep in mind this temperature may not exactly match what is listed on the manufacturer's data, so we may have to interpolate.)
- <u>Condition of air entering the indoor coil</u> – OEMs use either a dry bulb temperature, a wet bulb temperature, or may ask for some dry bulb – wet bulb combination.

Let's review our cooling target information:

---

16 *Manual J* requirements are always superseded by local code. If the code specifies other conditions, we are obligated by law to use those conditions.

17 The temperature of the air at the outdoor coil may be warmer than the outdoor temperature if the equipment is on a flat, black roof or near masonry walls, etc.

Table 10: Cooling Design Information				
	Total Load	Design Outdoor Temp	Target Airflow	Indoor Conditions
Cooling	26,107 Btu/h	91°F	1,025CFM	75°F , 50%Rh *
*75°F at 50% Rh.= 63°F wet bulb				

Now that we know what we are looking for, we can refer to OEM performance data and select a unit that will meet the cooling needs for Bob's house.

What You Always Wanted to Know About Manufacturer's Performance Data… But Were Afraid to Ask

We are going to look at two manufacturer's equipment. Various OEMs present their performance data in different ways. First we will look at the ABC Company.
When we examine the OEM expanded performance data we may see it laid out in different formats (Figures 13, and 14 next page offer two popular formats). There is no "best" way or "correct" way; these examples simply illustrate how two companies presented their equipment information. The HVAC contractor must take great care to read the whole table to prevent misreading the information and selecting the wrong equipment.

OEM Performance Data Differences (see two sample charts next page)

- The XYZ Company lists total capacity and sensible capacity for combinations of outdoor condensing temperature, airflow, and entering wet bulb. When reading these performance charts, one may assume the entering dry bulb temperature for Bob's house will be 75 °F (say ±1 °F). Conveniently that value is on the XYZ chart (next page) When another entering dry bulb temperature is selected by the designer then they must also apply an OEM provided adjustment procedure. Ensure you read all of the information on the performance data tables and apply it as necessary. Note: The OEM's data chart may or may not have the 75 °F value listed on it for entering dry bulb value.

- The ABC Company lists total and sensible capacities for combinations of outdoor condensing temperatures, entering wet bulb, and entering dry bulb, but only for one airflow rate. The ABC Company data also includes a table of adjustment factors for calculating capacities at different airflow rates.

- Also note that capacities may be listed as total Btu/h and sensible Btu/h, or as total Btu/h with a coil sensible heat ratio (equipment's SHR) value. These formats are equivalent (*sensible = total x SHR*).

ABC Performance Data						
Model 030 with Coil AC030 and Furnace FU060-36 @ 1,000 CFM						
OD Dry Bulb (F)	Indoor Entering Wet Bulb (F)	Total Capacity	Sensible Capacity at Entering Dry Bulb Temperature (F)			
			72	75	78	80
85	59	28,400	22,600	25,300	27,800	29,400
	63	29,900	18,800	21,600	24,300	26,100
	67	32,100	15,100	17,900	20,700	22,600
	71	34,700	11,400	14,200	17,000	18,900
95	59	27,300	22,200	24,900	27,400	28,300
	63	28,700	18,500	21,200	23,900	25,700
	67	30,800	14,700	17,500	20,400	22,200
	71	33,300	11,000	13,700	16,600	18,500
105	59	26,200	21,900	24,500	27,100	27,200
	63	27,600	18,100	20,900	23,600	25,400
	67	29,700	14,300	17,200	20,000	21,800
	71	32,100	10,600	13,300	16,200	18,100

OD Dry Bulb – Outdoor Dry Bulb, the outdoor temperature.

Correction Factors for other Airflows			
	Airflow	Total Capacity	Sensible Capacity
Low	875	0.98	0.93
High	1125	1.02	1.06
Multiply rated capacity data by factor.			

Figure 13: Representative OEM Performance Data – ABC 030

XYZ Air Conditioners – Detailed Cooling Capacities Model AC-30 with Coil AC-030							
Evaporator Air		Condenser Entering Air Temp – DB (F)					
CFM	EWB (F)	85		95		105	
		Capacity		Capacity		Capacity	
		Total	Sensible	Total	Sensible	Total	Sensible
875	72	34,610	18,190	33,100	17,620	31,520	17,020
	67	31,400	22,240	30,000	21,650	28,520	21,040
	63	28,620	26,290	27,350	25,680	26,020	25,040
	57	27,840	27,840	26,820	26,820	25,740	25,740
1000[18]	72	35,250	19,090	33,680	18,500	32,030	17,890
	67	31,990	23,660	30,530	23,060	29,000	22,440
	63	29,300	27,220	28,020	26,560	26,770	26,770
	57	29,020	29,020	27,930	27,930	26,780	26,780
1125	72	35,720	19,920	34,110	19,330	32,410	18,710
	67	32,430	25,010	30,930	24,410	29,360	23,780
	63	29,970	29,970	28,850	28,850	27,630	27,630
	57	30,000	30,000	28,850	28,850	27,640	27,640

Figure 14: Representative OEM Performance Data – XYZ AC-30

18 The SHR target value is 1,025 CFM (page 18) and the blower's capacity on Med-Lo fan speed was 995CFM. Using 1,000 CFM to evaluate the equipment capacity is the best choice. The difference between these values are minute (less than 2.5%) so no interpolation is needed. When we re-calculate the temperature rise for the furnace at this CFM, we see the airflow rate is still safe (procedure on page 15).

Previously, we discussed the *Manual J8* SHR and now we have a cooling coil or an equipment SHR. If the cooling equipment was specifically manufactured for Bob's house, the equipment's capacity would match the requirements. If the equipment sensible and latent capacities are equal or marginally exceed sensible and latent loads, Bob's house will be very comfortable during the summer.

The interpolated data for the competing air conditioners is summarized in Table 11[19]

Table 11: Comparing Design Criteria to Manufacturer's Data			
At 1000 CFM, 91°F† outdoor temperature, 63°F EWB††			
	Manual J8 Load	ABC Capacity	XYZ Capacity
Total Btu/h	26,107	29,180	28,532
Sensible Btu/h	21,423	21,360	26,824
Latent Btu/h†††	4,684	7,820	1,708

† The capacities listed were interpolated to 91°F from the listed capacities.

†† 63°F EWB is 75°F (DB) and 50%Rh.

††† The two manufacturers have only presented the data for total and sensible capacity. We must compute the difference to get the latent capacity.

Interpolation

The manufacturer's data shows 85°F and 95°F for condensing temperature, but our design temperature is 91°F. *Manual S*, N1-12 directs that if the design temperature is different than rated temperature, you must interpolate the data to get the more accurate information. Since 91°F is 4 degrees less than 95°F, we could simply use the 95°F data. For illustration purposes, the interpolation is performed here:

These calculations are for the **Total** Capacity of the XYZ model AC-030 air conditioner.

1. Determine difference between 95°F and 85°F	95 – 85 = 10
2. Determine difference between 91°F and 85°F	91 – 85 = 6
3. Divide the differences	6 ÷ 10 = 0.6
4. Determine difference between rated capacities at 85°F and the rated capacity at 95°F	29,300 – 28,020 = 1,280
5. Determine interpolation adjustment (Δ)	0.6 × 1,280 = 768
6. Subtract Δ from rated capacity of the higher (for 85°F) number	29,300 – 768 = 28,532 Btu/h
7. Perform a quick check to ensure the value is between the two source numbers	28,532 is between 29,300 and 28,020

Both air conditioning units deliver essentially the same total capacity. The difference is in the sensible and latent capacities. Although XYZ has more sensible capacity, ABC has about four times more latent capacity. Neither unit seems right for Bob's house. The ABC unit has too little sensible capacity and too much latent capacity. The XYZ unit is just the opposite.

19 *Manual S* (see summary in Appendix F) stipulates that the equipment's sensible capacity should not exceed the design capacity by more than 15% (i.e. a factor of 1.15). The interpolated sensible capacity of the ABC is within this range. But the sensible capacity of the XYZ model AC-30 air conditioner is 125.2% of the load. The XYZ model 030 is 25.2% over capacity so it would fail to meet the criteria to cool Bob's house.

In this case, the XYZ unit is immediately disqualified because the decidedly low latent capacity. Low latent capacity is a cardinal sin for HVAC systems in a humid or muggy climate. Fortunately for Bob, things are not as bad as they appear.

Manual S, §N 2-5, *Sizing Limits for AHRI-Certified Cooling-Only Equipment* discusses excess latent capacity, requirements for excess latent capacity usage to increase the sensible total. When we need more sensible capacity and less latent capacity then, per *Manual S* guidance, we add one half of the excess latent capacity to the sensible capacity.

Here is how the extra latent capacity is used. The excess latent capacity will make Bob's house a little drier than 50% Rh, perhaps 49% or 48%RH. At this lower humidity level, Bob will feel cooler though the thermostat may read higher than the design temperature of 75°F. Extra latent capacity is also desirable because indoor humidity tends to drift up during milder weather. Bob may feel a little too cool at 75°F if the house is less than 50% Rh, but that is easily corrected by turning up the thermostat until he finds a setting at which he feels comfortable. Good HVAC design helps Bob feel comfortable, and save money on his electric bill because his unit needs to run less.

The revised comparative data looks like this:

Table 12 Comparing Design Criteria to ABC's Adjusted Data			
1000 CFM, 91°F outdoor temperature, 63°F EWB and 75°F EDB			
	Manual J8 Load	ABC Capacity[20]	XYZ Capacity
Total Btu/h	26,107	29,180	28,532
Sensible Btu/h	21,423	22,928	26,824
Latent Btu/h	4,684	6,252	1,708

Now the ABC model 030 air conditioner with the AC030 evaporator coil is the clear winner.

It is fortunate that the manufacturer of the furnace and the air conditioner are one and the same. However, the air conditioner could be matched with a different manufacturer's furnace as long as the furnace blower capacity met the requirements of the air conditioner, in this case 1000 CFM. Remember, for heating and cooling, we selected a blower that could deliver about 1,000 CFM when working against 0.5 IWC or more of duct system resistance. The air conditioner (outdoor and indoor items) should always be an AHRI rated combination because it ensures the selected equipment is a rated match, capable of performing to set standards (see Figure 15 next page). In conclusion, the system that will meet Bob's heating and cooling requirements is the ABC 60-036 furnace and the ABC model 030 air conditioner matched with an AC030 evaporator coil.

[20] This is the "applied" sensible capacity which accounts for the excess latent capacity. Half of the excess latent capacity = equipment latent capacity – the latent load; excess latent = 7,820 Btu/h – 4,684 Btu/h = 3,236 Btu/h. Half of 3,136 Btu/h is 1,568 Btu/h. The sensible cooling capacity is 21,360. The sensible capacity with the added excess latent capacity (21,360 Btu/h + 1,568 Btu/h) is 22,928 Btu/h. This is an approximate value, not an actual one. The Total Capacity (Btu/h) remains the same, we merely allocate the sensible and latent capacity based on *Manual S* (see summary in Appendix F) and the cooling requirements for Bob's house.

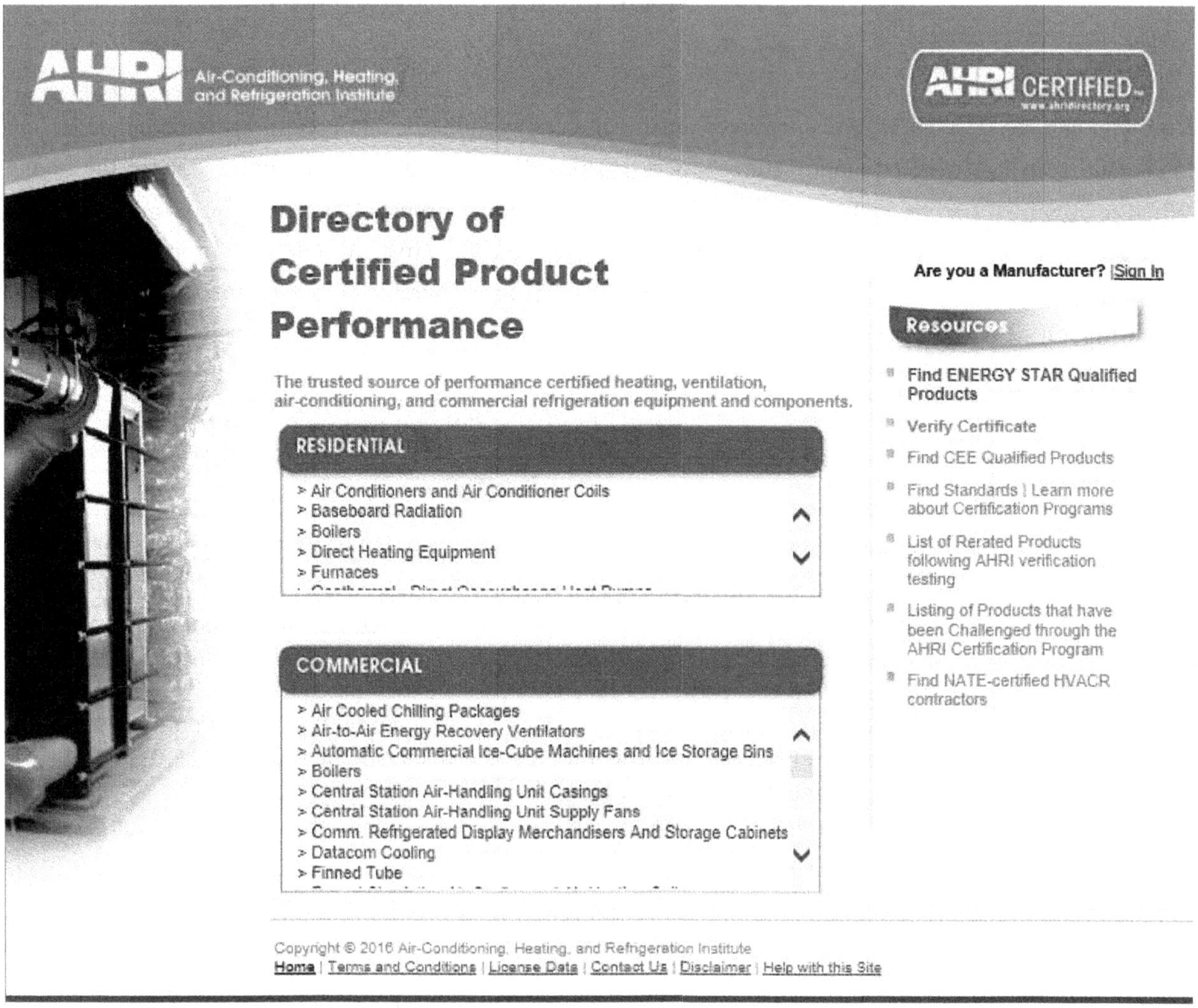

Figure 15: AHRI matching data is available on line at: www.ahridirectory.org

EQUIPMENT SELECTION SUMMARY

Now you can see why an HVAC contractor must use recommended design conditions to calculate design loads, and then use detailed manufacturer's performance data to find the right equipment for the home. When reviewing a permit application or evaluating a contractor's proposal, you may wish to check a few of the key items in the table below. Appendix D provides an illustrative form which you may find useful in evaluating an equipment selection. The example form in Appendix D is best interpreted when the supporting OEM performance data is attached.

Equipment Selection Checklist

Impact	Issue
High	Verify the equipment capacity is adequate for the heating and cooling loads.
	For heating, ensure the blower speed setting produces a safe temperature rise. For cooling, ensure the blower speed produces adequate sensible and latent capacity[1].
	Verify the manufacturers *expanded* performance data was used to select cooling equipment.
	Ensure the correct design parameters were used to extract the correct equipment capacity values from the manufacturer's performance data table[2]. (CFM, EDB, EWB, and Outdoor Temperature).
	Verify the equipment does not exceed the recommended *(Manual S)* limit for excess capacity. See Section *N2: Equipment Size Limits.*
Medium	Check the *Manual J8* sensible heat ratio (home's SHR) to determine the target cooling CFM.
	For heat pumps: balance point used to determine the amount of supplemental heat[3].
Low	SEER, HSPF and AFUE Ratings[4].
	Interpolate manufacturer's performance data if design parameters differ by more than 5 degrees.

1 For some applications, the same blower speed can be used for both heating and cooling, some homes may have heating and cooling equipment which may allow for different heating and cooling fan speeds.

2 The important thing is that the performance of the candidate equipment has been properly evaluated (as demonstrated by this document) and found to have adequate (but not excessive) sensible and latent capacity.

3 See Section 2 Manual S for a description of how to perform a balance point procedure.

4 Equipment properly selected and installed (complies with the minimum requirements of codes and the quality installation standard) has very good efficiency, though occupant comfort and total system efficiency are the relevant issues.

Figure 16: Equipment Selection Checklist

CHAPTER III
MANUAL D: PROPER AIR DISTRIBUTION

How Does the Air Get There From Here?

Air is the first word in air conditioning. Conditioned air has been heated or cooled, filtered, dehydrated or hydrated for circulation throughout the home. In winter, spring, summer and fall, we condition the air in our homes. If the duct system is designed by rules-of-thumb (sometimes endorsed by manufacturers and wholesalers who should know better), or by a "we have been doing it this way for twenty years" attitude, then:

- Bob's health and safety may be at risk,
- His utility cost could rise,
- The equipment performance could be deficient,
- The equipment could prematurely fail before its full, useful life, and
- Bob's family may not receive the comfort for which Bob paid so much money.

If Bob's house is going to be comfortable, the duct system must be designed to distribute the right amount of air (CFM) to each room and provide a low resistance return path from each room back to the furnace or air handler. Problems that arise when this is not done are discussed in Appendix C: Duct System Issues and Challenges.

The blower assembly in the furnace, selected in the previous chapter, can deliver a certain volume of air (995 CFM) when the fan is working against a certain amount of resistance (0.5IWC). The duct system must be designed to offer no more than 0.50 IWC of resistance so the blower can deliver 995 CFM. *Manual D* provides procedures to perform this very task.

Duct System Design and Duct Sizing Overview

Manual D uses the following worksheets, procedures, and tools:

- Effective Length Calculation Sheet
- Friction Rate Worksheet
- Duct Slide Rule
- Duct Sizing Worksheet

The *Manual D* procedures help the contractor to create a "duct blueprint", like the one in Figure 17 (next page), that a code official could inspect, a home owner could review, and an installation crew could use in the performance of their job.

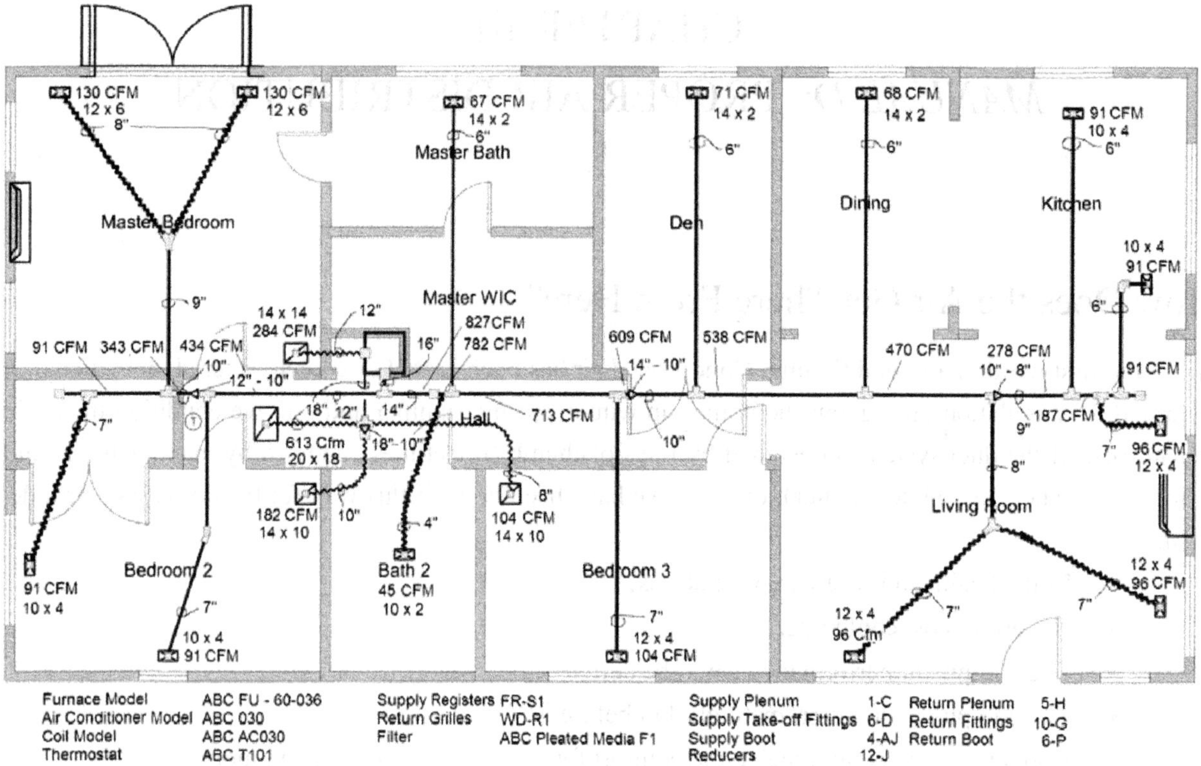

Figure 17: Final Layout Drawing

These are the steps we will follow to prepare a "plan":

1) Calculate each room's required CFM based on the *Manual J8* load calculation and the equipment's airflow capacity.
2) Sketch the duct blueprint as we expect it will be installed[21].
   a. Include notes showing approximate length of straight runs (or draw to scale).
   b. Show the CFM at each supply point and each return point.
   c. Calculate and show the CFM in each section of trunk duct (which changes every time air is taken from a supply trunk or added to a return trunk).
3) On the duct blueprint, identify each fitting required in the duct runs[22] (list the *Manual D*, Appendix 3 fitting code with an arrow to the fitting).
4) Gather copies (or transcribed notes) of manufacturer's blower performance data and pressure drop data for all air-side devices, such as cooling coil, boots, registers, grilles, etc.
5) Use the Effective Length Calculation Sheet to determine the longest supply and the longest return runs (uses steps 2 and 3 data).

---

[21] Note: Lines drawn on a floor plan show roughly where duct runs will be located, and are useful for estimating straight run lengths. HVAC installers will follow this plan, though they may route the duct through the available spaces and detour around columns, beams, joists and other obstructions. Coordination with the builder is crucial to prevent 'squeezing' a duct section through a tight space. Building officials should be very aware of squeezes and extreme deviations from the plan.

[22] Note: When working from plans, it may be impossible to anticipate every obstruction and problem, but a determined effort will produce adequate results. A ten percent error in the effective length estimate will not have a significant effect on system performance.

6) Complete the Friction Rate Worksheet[23] (use data from steps 3 and 4). The friction rate is the basis for all duct sizes and this is the most important step of the *Manual D* procedure. This step may be revisited to make adjustments, as necessary, before proceeding (i.e., increase or reduce blower speed, reduce length of critical circulation path, or find a fan with suitable blower performance).
7) Complete the Duct Sizing Worksheet (use step 6 data, a duct slide rule for the type of duct material used, CFM values for each section of the duct system, and the *Manual D* table for acceptable airway velocities.).
8) Add the duct sizes to the duct blueprint. Give copies to the code official, the home owner, and the installers. If the code officials do not require a copy, or no one has purchased the home, keep an extra copy in the job file.

Duct System Design and Duct Sizing Example

Our first step is to make sure every room receives the airflow required for proper heating and cooling. Partially complete the Duct Sizing Worksheet, perform the necessary calculations until the *Manual D* design CFM for the supply ducts, column six, is established for all rooms (see Figure 18 next page). This design supply CFM will serve as the basis for sizing the duct. The actual airflow may be less depending on whether Bob's House is balanced for cooling or heating. Also, the actual airflow may depend on Bob's personal comfort preferences. For either circumstance the duct will be large enough to meet the largest airflow required by the *Manual J* load calculation for the space.

Calculating Heating and Cooling Factors

Bob's house's heat loss (51,838 Btu/h) is more significant than the heat gain (21,423 Btu/h) – though in some climates this may be reversed. Heating and Cooling Factors (HF and CF) are used to determine which volume of load, and corresponding CFM, will be used to size the ducts.

Heating Factor = Blower CFM ÷ Manual J Heat Loss (Winter/Heating)

Cooling Factor = Blower CFM ÷ Manual J8 Sensible Heat Gain (Summer/Cooling)

In this case, the blower CFM is 995, the *Manual J8* heat loss is 51,838 Btu/h, and the *Manual J8* sensible cooling heat gain is 21,423 Btu/h, so:

HF = (995÷ 51,838) = 0.0192 *CF = (995÷21,423) = 0.0464*

These factors are multiplied by room load to determine room CFM.

Room Heating CFM = HF × Room Heating Load

Room Cooling CFM = CF x Room Sensible Load

The CFM values for Bob's Bedroom are:

Heating CFM = 10,202 Btu/h × 0.0192 = 196 Cooling CFM = 5,434 Btu/h × 0.0464 = 252

The higher of the two CFM will be considered the Design CFM for that room.

[23] Note: The *Manual D* procedure is not contingent on a worksheet. One could use scratch paper, but it does insist that we determine effective lengths with reasonable accuracy. Initially this takes time. A home may have 5 to 20 supply paths and possibly multiple return paths. With a little practice and experience in determining the path lengths, we can quickly investigate the lengths of a few 'long-run' candidates and ignore the other runs.

Duct Sizing Worksheet (cropped for use here)								
HF = Blower CFM ÷ *Manual J8* Heat Loss = (995 CFM) ÷ (51,838) = 0.0192								FR Value
CF = Blower CFM ÷ *Manual J8* Sensible Heat Gain = (995 CFM) ÷ (21,423) = 0.0464								
Supply Duct System								
Run - Trunk	**H – Btu/h**	**C –Btu/h**	**H – CFM**	**C – CFM**	**Design CFM**	**Round Size**	**Velocity**	**Final Size**
Master Bed	10,202	5,434	196	252	252			
Master Bath	3,306	685	63	32	63			
Walk-in-Closet	624	232	12	11	12			

Figure 18: Duct Sizing Worksheet – Design Airflow

The second step is to sketch the duct geometry.

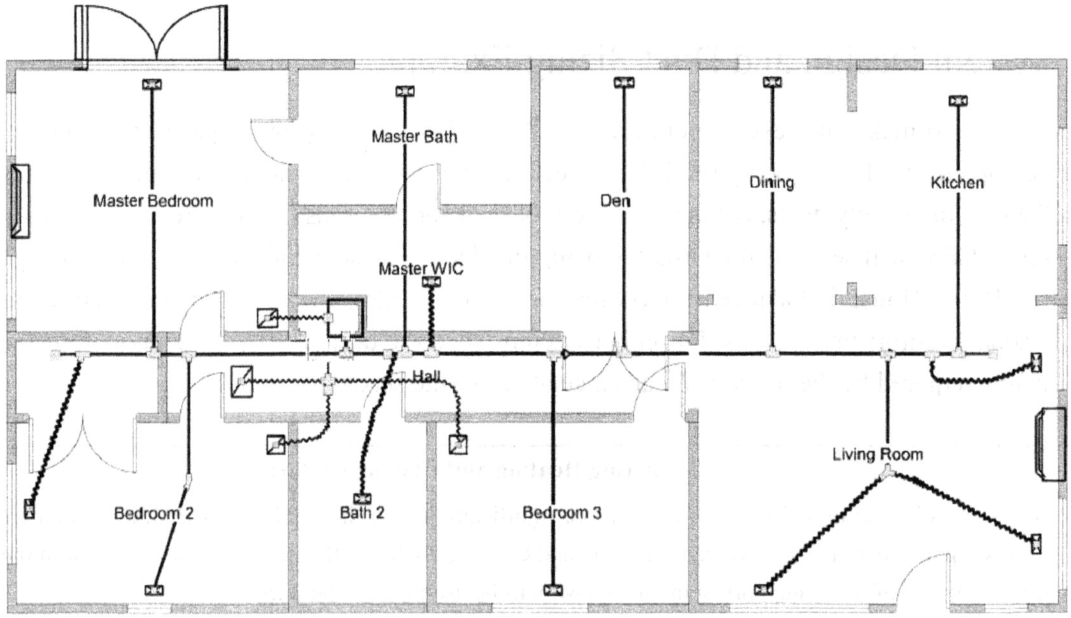

Figure 19: Sketch of Duct Layout on Floor Plan

We will continue our focus on Bob's master suite for the duct system design.

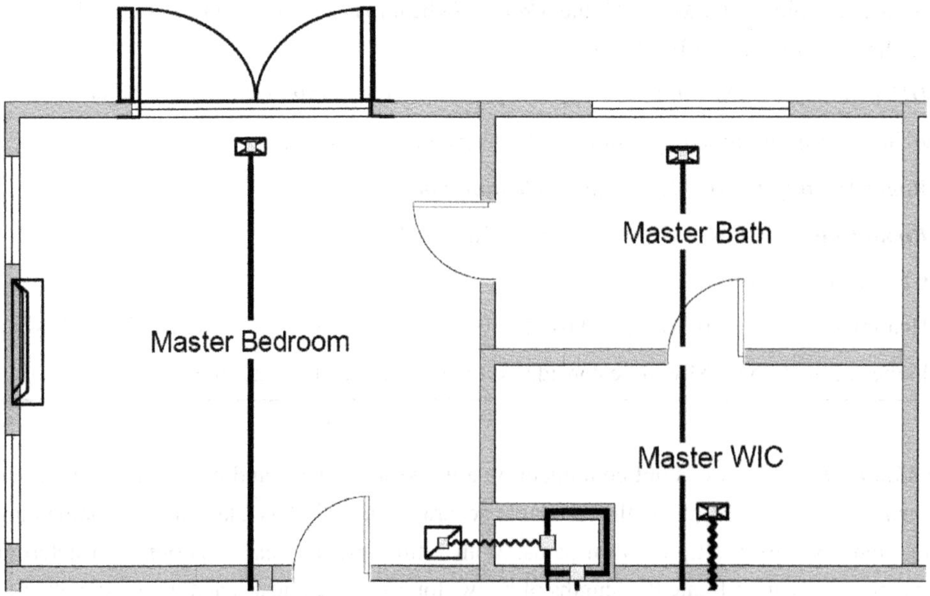

Figure 20: Duct Sizing Worksheet – Lead Airflow

After calculating the room heating and cooling airflow rates, we can examine if one or two branch ducts (and associated supply grilles) would best serve the room. For Bob's bedroom, two smaller grilles were selected rather than one large grille[24].

Duct Sizing Worksheet (cropped for use here)								
HF = Blower CFM ÷ *Manual J8* Heat Loss = (995 CFM) ÷ (51,838) = 0.0192							FR Value	
CF = Blower CFM ÷ *Manual J8* Sensible Heat Gain = (995 CFM) ÷ (21,423) = 0.0464								
Supply Duct System								
Run - Trunk	**H – Btu/h**	**C –Btu/h**	**H – CFM**	**C – CFM**	**Design CFM**	**Round Size**	**Velocity**	**Final Size**
Master Bed	10,202	5,434	196	252	252			
				Lead 1	126			
				Lead 2	126			
Master Bath	3,306	685	63	32	63			
Walk-in-Closet	624	232	12	10	12			

Figure 21: Duct Sizing Worksheet – Lead Airflow

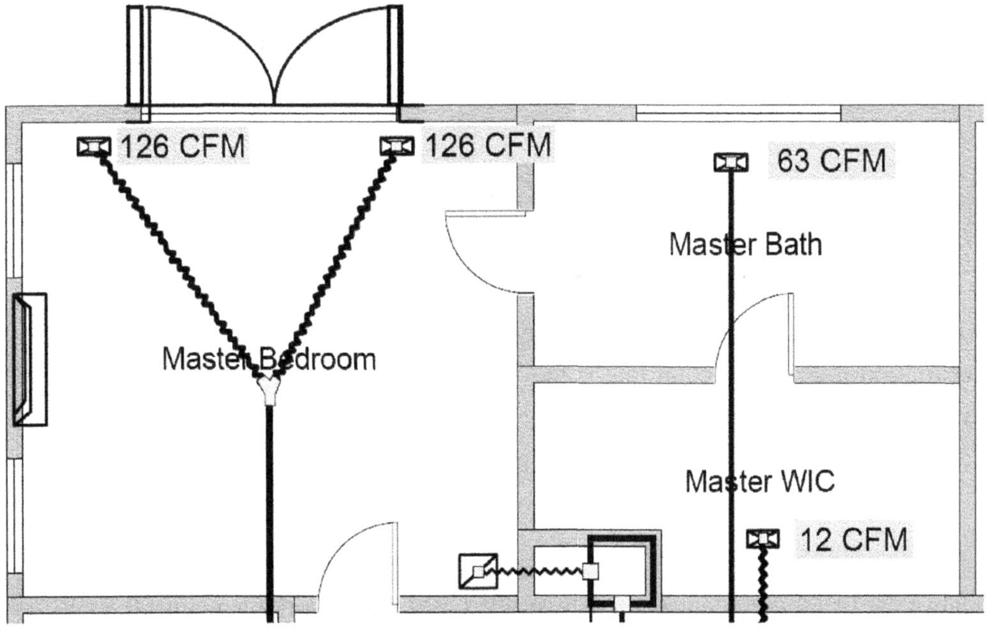

Figure 22: Bob's Suite Duct Layout on Floor Plan

Just like Bob's bedroom, most rooms typically have different airflow rates for heating and cooling. The design airflow rate is the larger of the two values[25]. This larger design duct size will ensure the duct(s) is sized for the larger design airflow, thus it will also be large enough for the smaller design flow rate.

On closer examination, we notice the airflow rate to the WIC is very low. From experience we know that moving small volumes of air is very difficult[26]. Residential register manufacturers have decided against

---

[24] ACCA *Manual T* explains room air mixing, duct outlet selection, and sizing, also briefly discussed in Chapter 4.

[25] This design CFM will serve as the basis for sizing the duct. This will ensure the duct will be large enough to meet the largest airflow required by the *Manual J* load calculation for this space.

[26] Installing supply registers in "any room I can walk in" does not always make good design sense. In some regions, adding this unnecessary cooling can cause condensation on the register or fail to provide even room temperatures. Evaluate the room's load, location, and use before arbitrarily adding a supply outlet.

fabricating a register or diffuser to adequately distribute less than about 30 CFM. The WIC is a room that is regularly used, but occupied for only short periods of time; it also has conditioned space surrounding it. Bob's HVAC Contractor recommends against conditioning the small room. When the facts are presented to him, Bob agrees.

The HVAC Contractor still accounts for the small load and airflow. Figure 23 shows Bob's revised master suite duct sketch after the WIC load (12 CFM) is redistributed.

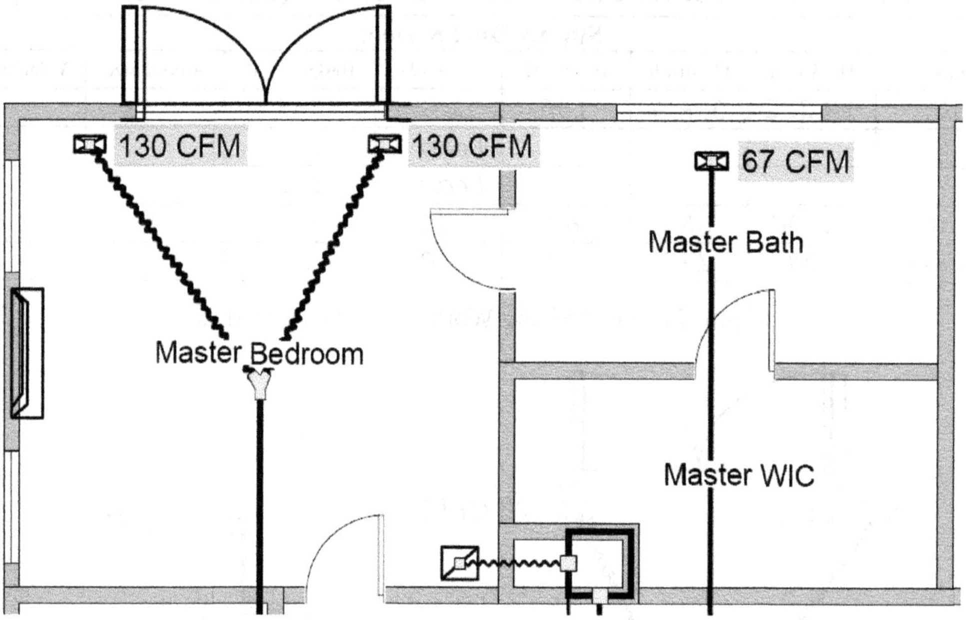

Figure 23: Bob's Suite with Adjusted Air Flow

Now that we know the general path that the duct system will follow and the rooms design airflow, we can calculate the friction rate. The friction rate is the "yard stick" we use to size the duct system's airways. The friction rate is based on three factors:

- The blower airflow capacity from the manufacturer's performance table
- The resistance from airside devices (i.e., high efficiency filters, A/C coils, etc.)
- The resistance from the duct

Figure 24 shows the FU60-036 blower can deliver 995 CFM when working against 0.5 IWC of resistance from everything in the duct system.

Air Delivery – CFM (With Filter)								
Unit Size	Speed	External Static Pressure (inches water column)						
		0.1	0.2	0.3	0.4	0.5	0.6	0.7
FU60 - 024	High	1075	1040	995	945	895	840	760
	Med – Hi	950	925	895	845	795	740	660
	Med – Lo	850	825	780	740	685	635	560
	Low	740	700	650	620	565	515	455
FU60 - 036	High	1470	1415	1400	1285	1[illegible]15	1120	995
	Med – Hi	1315	1280	1235	1298	1[illegible]15	1035	930
	Med – Lo	1125	1110	1085	1045	995	915	830
	Low	930	9256	910	850	830	770	705
Note: Airflow for FU60-036 based on a dry coil at 120v with factory approved filter (0.06IWC).								

Figure 24: Sample OEM Blower Performance Data

Next, we determine how much resistance is produced by the airside devices[27] and the duct system, remembering that we can only use 0.50 IWC of external static pressure from Figure 24. The resistance from the air-side devices comes from the manufacturer's performance data and *Manual D* default values.

The ABC FU60-036 can produce: 995 CFM @ **0.50 IWC** Total External Static Pressure

The air conditioning coil produces (A)	0.20 IWC resistance (wet)
An accessory filter produces (B)	0.12 IWC resistance (clean)
Deduction for removing OEM filter[28]	-0.06 IWC resistance (clean)
Our supply register (C) has[29]	0.03 IWC resistance
Our return grille (D) has	0.03 IWC resistance
The balancing damper (E) has	0.03 IWC resistance
The Total Resistance For All Air-Side Devices is:	**0.35 IWC** Total Resistance
The Available Static Pressure (ASP) is:	**0.15 IWC** ASP[30]

0.15 IWC is the available pressure remaining from the fan's capacity for the duct system.

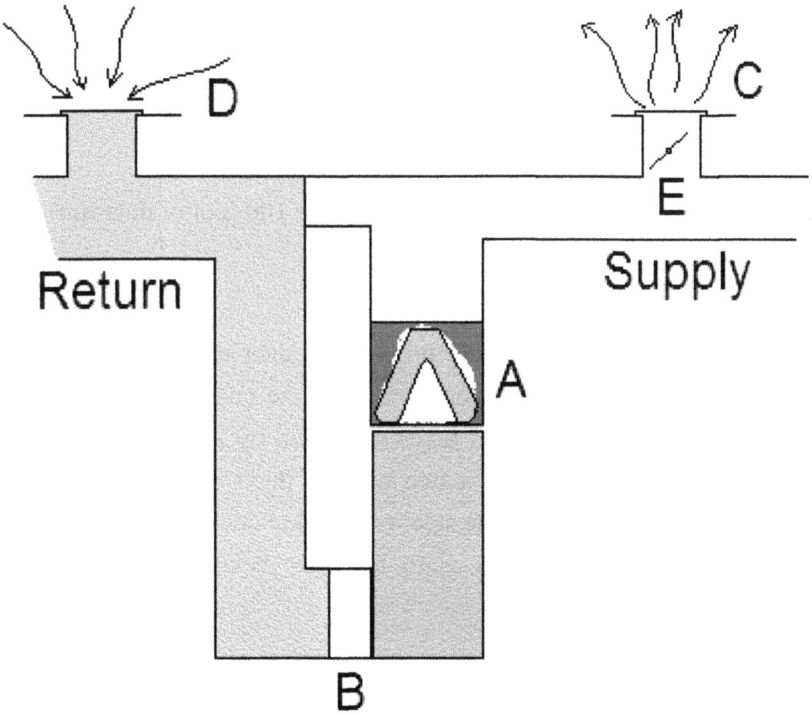

Figure 25: Examples of Duct System Resistance Devices

After subtracting the total device resistance from the total ESP the ASP is determined to be 0.15 IWC. This is the pressure available to move the air through the critical (longest) circulation path of the duct's

[27] The difference between the ESP and the total resistance (pressure measured in IWC) from the airside devices is the ASP.

[28] The furnace airflow was rated with factory filter in it (see Note at bottom of Figure 24). The factory filter will be removed to add a high-efficiency filter, so we subtract the resistance for the factory filter (0.06IWC), and add the resistance for the high-efficiency filter (0.12).

[29] *Manual D* provides default values for grilles, registers and balancing dampers. We only need to count these items once because this represents one circuit: in the return grill, through the return duct, through the unit, through the supply duct, and finally through the supply grille.

[30] ASP = Total ESP (0.50) - Total Resistance For All Air-Side Devices (0.35) = 0.15 IWC ASP.

straight runs and fittings. The straight lengths of duct are easy to measure, however, duct runs also have elbows, branch take-offs, terminal air boots, plenums and various other fittings.

The airflow resistance for common residential fittings is provided by *Manual D*, Appendix 3. The resistance of duct fittings is equated to a length of straight duct that would produce the same (or equivalent) resistance. For example, an eight inch long reducer may have an equivalent length (EL) of 20 feet because the pressure drop through the fitting equals the pressure drop through 20 feet of straight duct. To calculate the total effective length (TEL) of a duct run:

1. Find the fittings in Appendix 3 and transcribe the EL values to the duct blueprint.
2. Measure or scale the lengths of straight duct, and post on the duct blueprint.
3. Calculate the effective length of the longest supply and longest return run[31].
4. Add the longest supply effective length to the longest return effective length to get the total effective length for the duct system.

For Bob's house, the duct system has a TEL of 165 feet. Several detailed examples for determining effective lengths are worked out in *Manual D.*

For Bob's house the friction rate is 0.09 IWC[32] per 100 Foot of duct

$$\boldsymbol{FR = (ASP \times 100) \div TEL}$$

$$FR = (0.15 \times 100) \div 165 = \boldsymbol{0.09}$$

Where:

FR = Friction Rate (inches water column pressure drop per 100 foot of duct run)

ASP = Available Static Pressure

TEL = Total Effective Length

The calculated FR value is critical to selecting the proper size duct to carry the necessary air. The methods and materials used to fabricate duct systems vary. Some contractors use smooth galvanized sheet metal for their ducts. Others use thin vinyl tubes supported by a flexible steel wire frame, also known as "flex". The resistance to airflow of these two types of ducts is very different. Sheet metal is very rigid and smooth, and has low resistance. The steel ribs in flex duct are like speed bumps, giving it higher resistance. There are different friction rate charts for sheet metal, for flexible duct, and for duct board (see *Manual D*, Appendix 2[33]).

---

31 For any duct system, there are many supply runs and one or more return runs. We are only interested in determining the longest supply run and the longest return run. Inexperienced designers must use the "brute force" method (calculate the lengths of all runs and look to see which is the longest). This method is also ideal for computer software, but as we gain experience we will begin to notice the attributes of long runs. Only a few runs have the attributes of "the longest run", so our TEL calculations will be reduced to only a few candidates.

32 Many contractors use a "rule of thumb" that the friction rate is 0.10 IWC. However, as you can see, the actual FR is highly dependant on the ASP, a function of the equipment and additional components selected, and the TEL, a function of the duct system design.

33 Compare the airflow rates for a sheet metal and a flex duct of the same diameter. If we use the 0.09 friction rate and the friction rate charts for metal and flex duct we will see a big difference in airflow capacity. For example:

- A six inch round **sheet metal** pipe at 0.09 friction rate can carry about 100 CFM of airflow.
- A six inch round **flex** duct at 0.09 friction rate can only accommodate 80 CFM of airflow.

That 20% difference can be the difference between a comfortable and an uncomfortable room.

The calculated friction rate is applied to the Duct Sizing Worksheet and now the duct sizes can be determined.

Duct Sizing Worksheet (cropped for use here)								
HF = Blower CFM ÷ *Manual J8* Heat Loss = (995 CFM) ÷ (51,838) = 0.0192								FR Value
CF = Blower CFM ÷ *Manual J8* Sensible Heat Gain = (995 CFM) ÷ (21,423) = 0.0464								0.09
Supply Duct System								
Run - Trunk	**H – Btu/h**	**C –Btu/h**	**H – CFM**	**C – CFM**	**Design CFM**	**Round Size**	**Velocity**	**Final Size**
Master Bed	10,202	5,434	198	252	252			
	CFM redistributed for walk-in closet			Lead 1	~~126~~ 130			
				Lead 2	~~126~~ 130			
Master Bath	3,306	685	63	32	~~63~~ 67			

Figure 26: Duct Sizing Worksheet – Friction Rate

At the bottom of the Friction Rate Worksheet is a chart. This chart graphically illustrates the mathematics calculated above. In addition graphically demonstrating the friction rate, it also depicts a "wedge" of design friction rates. Selecting a friction rate outside of this wedge may lead to problems with duct velocity.

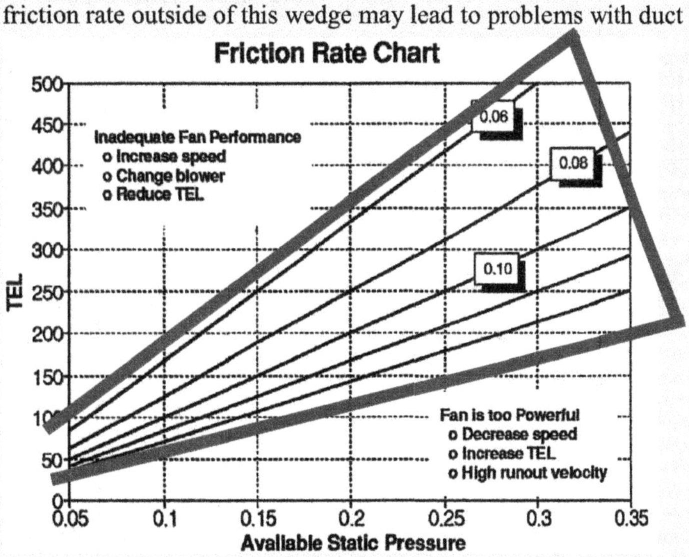

Figure 27: Friction Rate Worksheet Wedge

Some branch ducts for Bob's master suite are flexible duct as denoted by the squiggly line on the duct blueprint. We can use the friction rate charts from *Manual D* Appendix 2 or we can use a duct calculator which is just as accurate and, with practice, much faster.

Recent research from Texas A&M University has demonstrated that flexible wire helix duct (flex duct) installation, compression[34], and sag[35], greatly affects its resistance, and therefore its ability to move air. *Manual D* procedures and duct sizing slide rule are based on proper flex duct installations (no sag & no more than 4% compression)

[34] Compression is the difference between a flex duct that is fully extended (pulled taut at both ends) versus when flex that has "relaxed", or contracted, or compressed after being stretched. For example, a duct that is 25' long when pulled tight might compress to 24' when released, compression ratio 1 ft ÷ 25 ft = 0.04 or 4% compression.

[35] Sag is caused a lack of support from poor installation; the duct will droop rather than be straight and tight. If a flex duct has 4% or less compression and is properly supported every five feet, then sag will be negligible.

Using the *Manual D* Appendix 2 chart for flexible ducts, we size the supply run outs (or leads) for Bob's bedroom. 130 CFM needs a 7.5" duct, the closest standard size is an 8" duct. So, on the Duct Sizing Worksheet in Bob's master bedroom, we post an 8 inch duct, this will later be transcribed to the blueprints or duct sketch (Supply ducts Figure 29, Return ducts Figure 31).

Duct Sizing Worksheet
(cropped for use here)

	FR Value
HF = Blower CFM ÷ *Manual J8* Heat Loss = (995 CFM) ÷ (51,838) = 0.0192	
CF = Blower CFM ÷ *Manual J8* Sensible Heat Gain = (995 CFM) ÷ (21,423) = 0.0464	0.09

Supply Duct System

Run - Trunk	H – Btu/h	C –Btu/h	H – CFM	C – CFM	Design CFM	Round Size	Velocity	Final Size
Master Bed	10,202	5,434	198	252	252			
	CFM redistributed for walk-in closet			Lead 1	~~126~~ 130	8"		
				Lead 2	~~126~~ 130	8"		
Master Bath	3,306	685	63	32	~~63~~ 67	6"		

Figure 28: Duct Sizing Worksheet – Duct Sizes

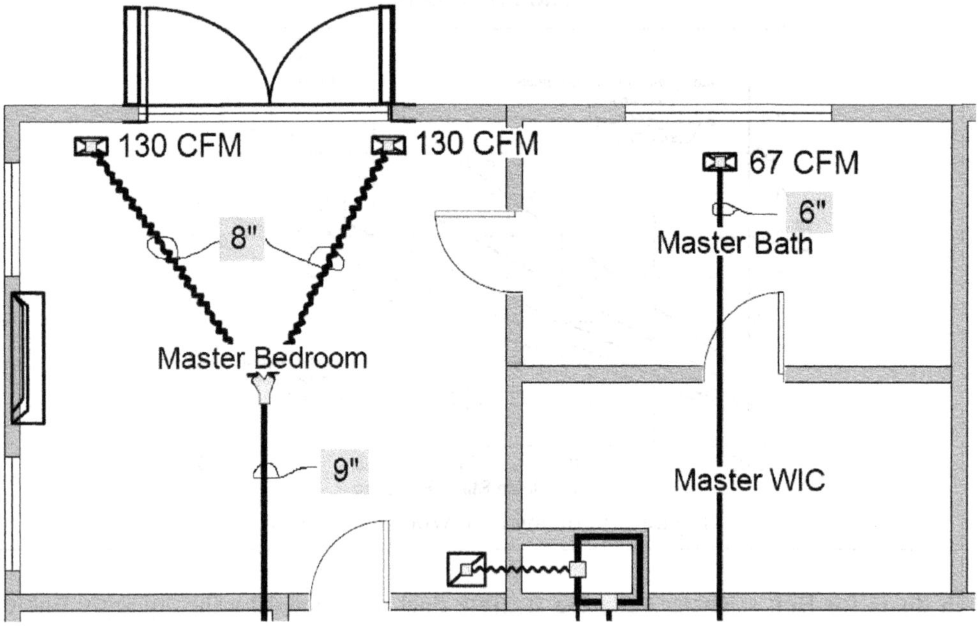

Figure 29: Bob's Suite with Duct Sizes

An 8" duct does not mean an automatic 130 CFM. Bob's ducts need dampers to adjust or balance the air flow to meet the design. In Bob's master bedroom, one damper could be installed in the 9" branch take-off fitting and both supply ducts would be adjusted by one damper, the installer would adjust the damper until the air from both supply registers equaled 260 CFM.

The return duct size follows the same basic rules. Though the total design supply CFM is not used to size the return duct for Bob's master suite. The return duct size for Bob's master suite is based on the larger of the total *heating* supply airflow, <u>or</u> the total *cooling* supply airflow. In Bob's master suite the total heating airflow is 261 CFM, and the total cooling airflow is 327 CFM, we will size the return for 327 CFM.

We see from the Duct Sizing Worksheet (Figure 30) that the design airflow for the return will be based on the cooling CFM, 327 CFM. The duct sketch (Figure 31) illustrates the associated duct size for the return trunk based on the friction rate.

Duct Sizing Worksheet (cropped for use here)							
HF = Blower CFM ÷ *Manual J8* Heat Loss = (995 CFM) ÷ (51,838) = 0.0192						FR Value	
CF = Blower CFM ÷ *Manual J8* Sensible Heat Gain = (995 CFM) ÷ (21,423) = 0.0464						0.09	
Return **Duct System**							
Run - Trunk	Associated Supply Runs	H – CFM	C – CFM	Design CFM	Round Size	Velocity	Final Size
Master Suite	S1, S2, S3	261	327	327	10"		

Figure 30: Duct Sizing Worksheet – Return Duct Sizing

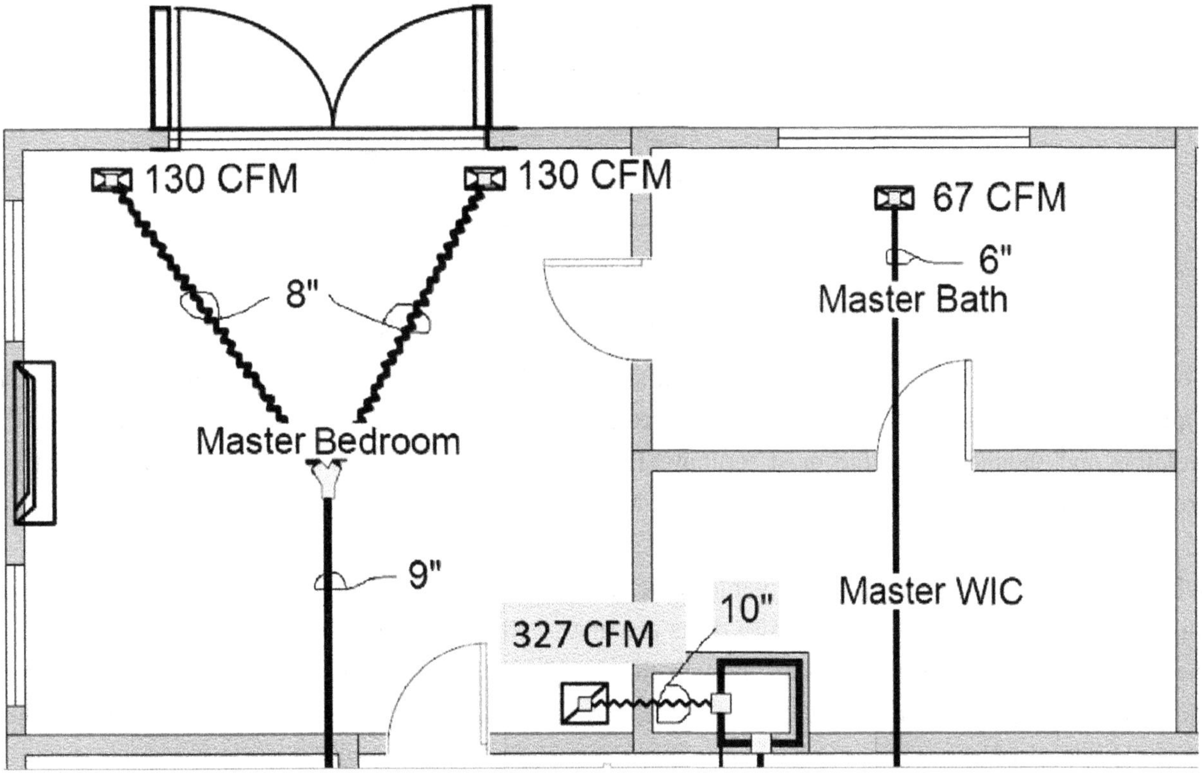

Figure 31: Bob's Suite with Return Duct Sizes

The returns design CFM, either heating or cooling, is used to size the duct. Ducts sized to accommodate either the heating or cooling CFM will provide the flexibility to carry larger of the two airflows. Homes which use a zoning system, variable volume airflow, and staged or variable refrigerant flow could provide different airflow rates for winter or summer. A constant volume system, like the one in Bob's house does not have that level of control.

What's Your Hurry?

After ducts are sized using the friction rate, the designer must ensure the duct velocities do not exceed the recommended limits. If air velocity is too high, the duct may produce objectionable noise. When *Manual D* procedures and friction rates remain within the "wedge" found on the Friction Rate Worksheet (Figure 32), the *supply* duct velocities will usually be acceptable.

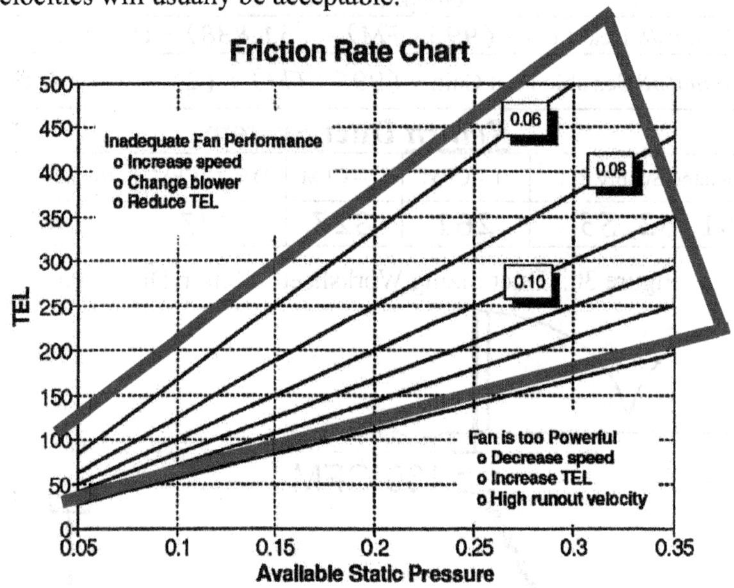

Figure 32: Friction Rate Worksheet Wedge

Contractors should check air velocities for each supply and return branch, and each unique[36] section of supply and return trunk. In general, trunk and return branch ducts are usually the biggest "speed limit" violators and should receive the most scrutiny. The final design for Bob's bedroom ***return*** air duct is presented here. A 12" flex duct will carry 284 CFM along at a quiet 365 FPM.

When air velocity is too high, unacceptable noise can be produced, return ducts are more susceptible to this than supply ducts. Therefore, return ducts are first sized for appropriate resistance to airflow using the friction rate, and then air velocity through the duct is considered. The duct size is evaluated and increased if necessary. This increase may "over-size" the duct as far as resistance is concerned; the larger, quieter duct provides a reason for installing balancing dampers in return ducts too.

Bob's return air duct is carrying a volume of 327 CFM through a 10" flex duct at velocity of 610 FPM.

Duct Sizing Worksheet							
HF = Blower CFM ÷ *Manual J8* Heat Loss = (995 CFM) ÷ (51,838) = 0.0192						FR Value	
CF = Blower CFM ÷ *Manual J8* Sensible Heat Gain = (995 CFM) ÷ (21,423) = 0.0464						0.09	
Return **Duct System**							
Run - Trunk	Associated Supply Runs	H – CFM	C – CFM	Design CFM	Round Size	Velocity	Final Size
Master Suite	S1, S2, S3	261	327	327	10"	610	

Figure 33: Duct Sizing Worksheet – Return Duct Velocity

[36] *"Unique"* being a change in CFM value and/or airway size.

When consulting the recommended velocity chart, Figure 34, we see that the recommended velocity for a flex return branch is 400 FPM; the maximum velocity is 700 FPM.

Recommended Velocity (FPM)								
Type of Run	Supply				Return			
	Recommended		Maximum		Recommended		Maximum	
	Rigid	Flex	Rigid	Flex	Rigid	Flex	Rigid	Flex
Trunk Ducts	700	600	900	700	600	600	700	700
Branch Ducts	600	600	900	700	400	400	700	700
Manual D, Table 3-1								

Figure 34: Recommended Duct Velocity Chart

610 FPM is acceptable, but to comply with the recommended velocity and give Bob a quieter bedroom we will re-size the return duct.

Duct Sizing Worksheet (cropped for use here)							
HF = Blower CFM ÷ *Manual J8* Heat Loss = (995 CFM) ÷ (51,838) = 0.0192							FR Value 0.09
CF = Blower CFM ÷ *Manual J8* Sensible Heat Gain = (995 CFM) ÷ (21,423) = 0.0464							
Return **Duct System**							
Run - Trunk	Associated Supply Runs	H – CFM	C – CFM	Design CFM	Round Size	Velocity	Final Size
Master Suite	S1, S2, S3	261	327	327	10"	610	12"

Figure 35: Duct Sizing Worksheet – Re- Sizing Return Duct

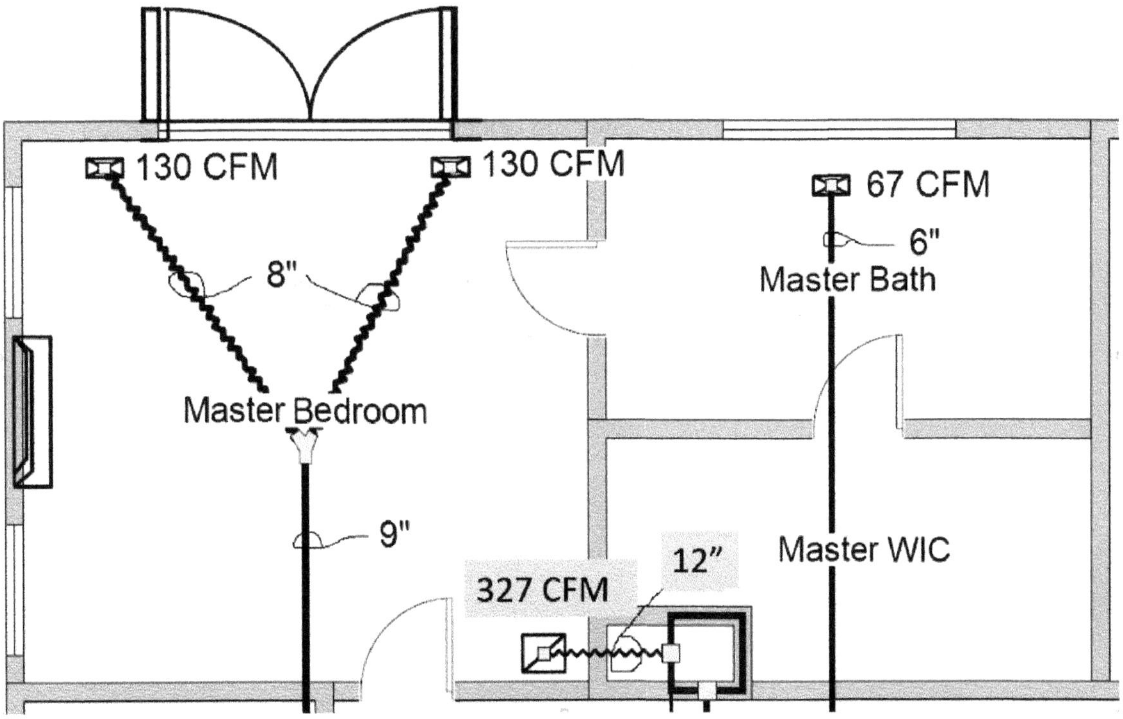

Figure 36: Bob's Suite with Return Duct Sizes

Once again, a 12" flex duct does not automatically equal 327 CFM, a damper should be installed, the airflow measured, and the damper adjusted until the proper airflow is achieved.

AIR DISTRIBUTION SUMMARY

Air is the first word in air conditioning. Air flows through duct systems into homes with central heating and air conditioning. The duct system must be designed to carry the correct volumes of air, or the equipment will fail to function effectively, be energy efficient, and the occupant's comfort is compromised or sacrificed. Duct system design is time consuming and meticulous, but it is very necessary for effective comfort control. The key elements to the design of a good air distribution system are listed below. The impact each issue has on the outcome of the duct design is also ranked by importance and effect on the duct system design. Each element is connected to the other, so great care must be taken to avoid making a mistake in one area which will affect the whole design.

Air Distribution Design Checklist

Impact	Issues
High	Manufacturer's data – blower fan performance
	Calculation of available static pressure (ASP)
	Total effective length (TEL) and room CFM values
	Design friction rate value
	Adequate return air paths for all rooms
	Dampers installed, air flow balanced
Medium	Load calculation data transfer - CFM for each room
	Duct airway flow rates and velocities
	Air distribution system layout (with duct sizes)
Low	Pressure drop produced by terminal devices
	Location of return (regarding high or low position)

Figure 37: Air Distribution Checklist

CHAPTER IV
MANUAL T: SELECTING THE CORRECT GRILLES AND REGISTERS

Sprint to the Finish Line

When you have reached this stage in the process, you need to finish well. The load has been carefully calculated. The equipment has been selected using the OEM data. The duct system has been designed to deliver the correct amount of air to each room. Now is the time to ensure all this hard work will be effective. We turn our attention to the grilles, registers, or diffusers (terminal devices).

Terminal devices are just as important as the thermostat, air conditioner, furnace, or duct system. Poor terminal devices can make a well-designed, expertly installed, perfectly balanced heating and cooling system appear cheap, amateurish, and inadequate. Terminal devices must by the correct type, style, and size and in appropriate locations[37]. The wrong terminal devices can:

- Create drafts that cause Bob discomfort when he sleeps, eats, or watches a ball game.
- Whistle and hum. Noisy grilles are aggravating, and in the extreme can interfere with conversations, TV, or radio programs.
- Cause stagnant and stratified air. The room may feel stuffy because the air moves too slowly and does not mix well or circulate.[38]

Manual D provides basic guidance for selecting terminal devices. However, ACCA *Manual T* goes into much more detail.

To select terminal devices, consider their location in the room and the affect you want them to have on the room air circulation. Different seasons and different register or diffuser locations will have impact the room differently. We want to select registers or diffusers with certain characteristics depending on the season, the predominant air conditioning mode, or the predominant architectural consideration. These considerations are the main emphasis of our register selection.

- In the winter we want to ensure we select:
  - Ceiling supply registers with enough vertical "*throw*" to maximize mixing warm buoyant supply air with room air before the warm air floats up and stratifies at

[37] There is an industrywide misconception that oversized ducts have an undesirable effect on room air motion. This is incorrect, because room air motion is greatly controlled (or not) by the performance characteristics of the supply outlet and the velocity of the air flowing through the face of the grille or register.

[38] If the conditioned air is not properly mixed with room air, the temperature at knee level may be cool while the temperature at shoulder level may be warm.

the ceiling. Rooms with very high ceilings need to pay special attention to this performance factor.
    - Floor supply registers with enough horizontal "*spread*" to ensure warm buoyant supply air mixes well with room air as it rises to the ceiling.
- In the summer we want:
    - Ceiling supply registers with enough horizontal "*spread*" to ensure the cool supply air mixes with the room air before it settles near the floor.
    - Floor supply registers with enough vertical "*throw*" to ensure cool air gently hits the ceiling, and tumbles to mix the cool conditioned air with warmer room air.
- Low and high side wall supply side wall registers should be avoided because it is difficult to mix room air without creating unwanted drafts in the occupied space.
- All terminal devices must balance the air volume delivered and the velocity at which it is delivered. Each home's floor plan, duct design, and architectural circumstances will guide the selection. Since more velocity means more throw and more noise, we may have to investigate the performance of more than one device to obtain the desired result. All supply terminals are not created equal.

Like equipment selection, the key to terminal device selection lies in the manufacturer's engineering data. This data consists of tables (Figures 38 and 39) that correlate:

- The size of the register, grille or diffuser,
- The flow rate (CFM) passing through the device,
- The throw and spread of the air to address cooling and heating performance,
- The average velocity (Table 14 provides the recommended velocities for supply and return terminal devices.),
- The resistance (IWC) the device creates at a given CFM.

Table 14: Recommended Velocities (FPM) for Terminal Devices			
	Supply		Return
	Recommended	Maximum	Maximum
Supply Outlet Face Velocity	Size for Throw	700	
Return Grille Face Velocity			500
Filter Grille Face Velocity			300
Portion of table found in *Manual D*, Section 1: Table A1-1			

As was the case for equipment performance data, air terminal manufacturers use different formats to present their products performance characteristics. Figures 38 and 39 (next page) provide one sample format.

For Bob's bedroom, we need 130 CFM for each floor register. The bedroom has an eight foot ceiling, so we look for a register with an eight foot throw. We also want a register that is quiet, so we keep our eye on face velocity. The FR-S1, 12 x 6 looks like a good fit. At 125 CFM, it has almost eight feet of throw, and the face velocity is approximately 500 feet per minute. If we interpolate for the increased volume, the velocity is 525 FPM, the throw is closer to 8 ft, and the pressure drop is 0.018 IWC, so the device produces less flow resistance than the *Manual D* worksheet default value of 0.030 IWC used in Section 2 of this guide to determine the available static pressure (ASP).

Floor Register Model FR-S1						
Face Velocity		400	500	600	700	800
Pressure Loss		.010	.016	.022	.031	.040
14 x 2	CFM	45	60	75	90	105
	Throw	4.5	6	8	8	9.5
12 x 6	CFM	110	125	145	170	200
	Throw	6	7.5	9	10	12
14 x 6	CFM	130	165	200	230	265
	Throw	6.5	8	9.5	11	13

Figure 38: Sample OEM Performance Data Supply Registers

Some grille manufacturers use a removable core; the core adjusts airflow for high throw or wide spread. This allows the home owner to optimize airflow seasonally by simply adjusting the core.

Return grilles are very similar. For Bob's bedroom suite the maximum total return flow volume is 284 CFM. The WD-R1 14 x 14 looks like the best match. At 284 CFM it has a face velocity of about 305 FPM. This will be a nice quiet grille, and a pressure drop of 0.010 is compatible with the *Manual D* default value of 0.030 IWC. Since it is a smaller value it will not increase the ASP used for sizing the duct.

Return Grille Model WD-R1						
Face Velocity		300	400	500	600	700
14 x 8	CFM	158	211	264	316	369
	Pressure Loss	0.010	0.018	0.029	0.041	0.056
14 x 10	CFM	198	265	331	397	463
	Pressure Loss	0.010	0.018	0.029	0.041	0.056
14 x 12	CFM	239	318	398	477	557
	Pressure Loss	0.010	0.018	0.029	0.041	0.056
14 x 14	CFM	279	372	465	558	651
	Pressure Loss	0.010	0.018	0.029	0.041	0.056
14 x 18	CFM	360	480	600	720	840
	Pressure Loss	0.010	0.018	0.029	0.041	0.056
16 x 18	CFM	460	580	700	820	940
	Pressure Loss	0.010	0.018	0.029	0.041	0.056

Figure 39: Sample OEM Performance Data Return Grilles

After selecting all of the grilles and registers, we can complete the air distribution system design by adding all duct system details to the plan.

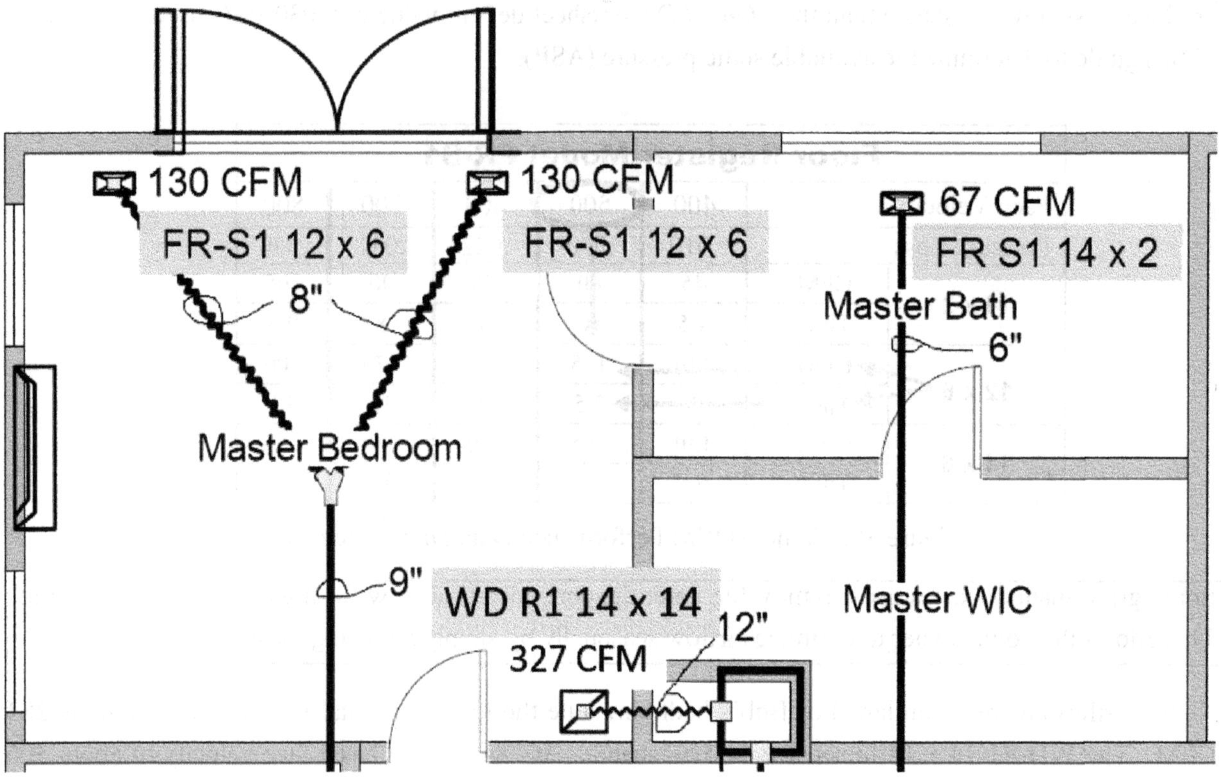

Figure 40: Bob's Suite with Register and Grille Sizes

As shown here, we now have a mechanical plan based on recognized good HVAC design practices and the model building codes.

On the next page, we see a full plan that could be used by the installation crew. This should provide some amount of confidence and satisfaction for Bob, and it is something an inspecting code official can use when evaluating the installation.

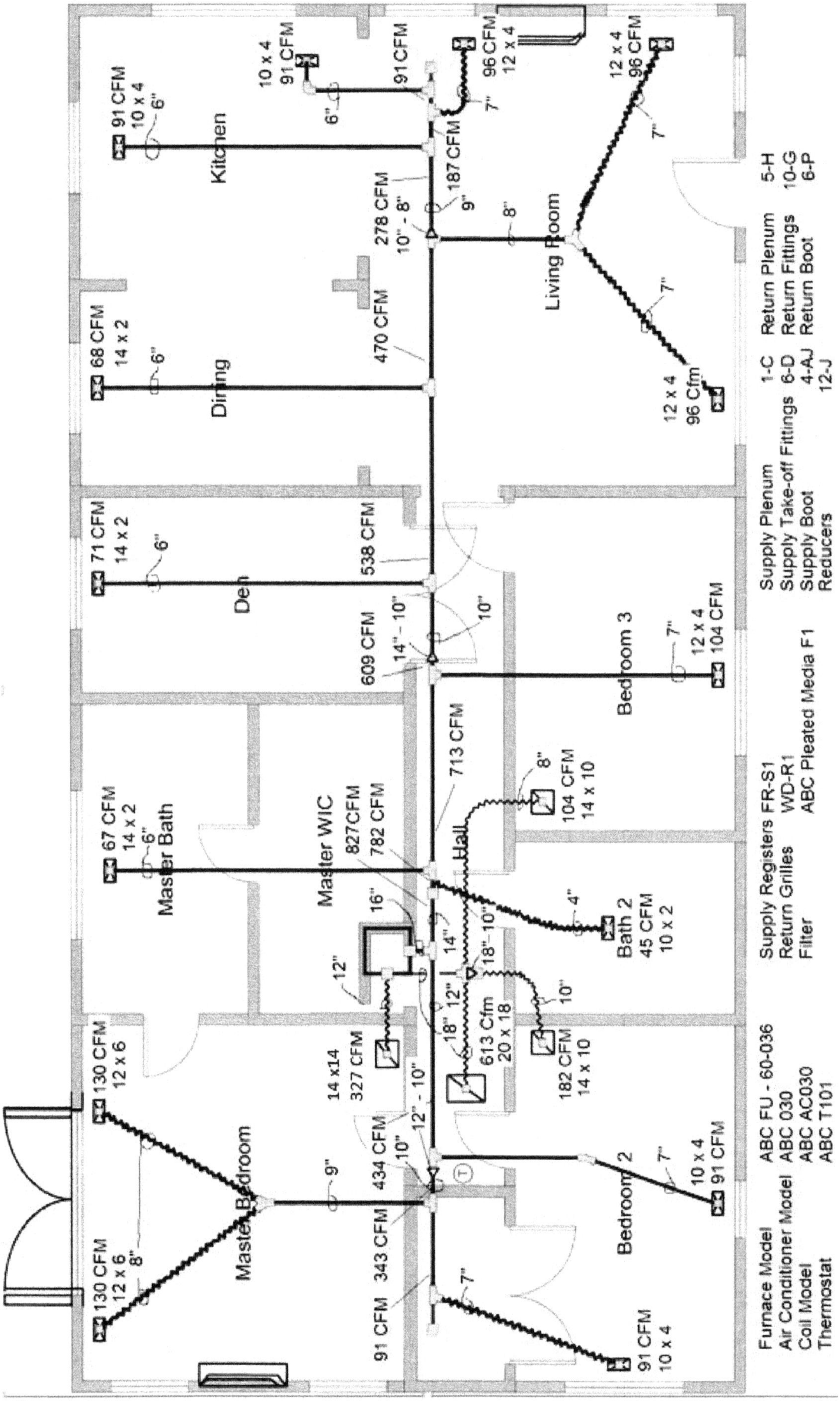

Figure 41: Enlarged view of Bob's Layout

Bob's House

SELECTING GRILLES AND REGISTERS SUMMARY

Grilles and registers are usually the least expensive part of the HVAC system. A person rarely walks into a room and immediately looks for the ceiling diffusers or return grilles unless he or she is an HVAC geek. Great care can be taken to perform a rigorous load calculation, the perfect unit can be selected from the masses, and the duct system can be meticulously refined, BUT if the registers rattle, whistle, or are drafty, it may be considered to be a cheap or shoddy HVAC system. Grilles, registers and diffusers are a very small but vital portion of the HVAC system, and should be selected and examined with the same scrutiny as the other portions of the design process. The key elements to the design of a good air distribution system are indicated below:

Air Distribution Design Checklist	
Impact	Issues
High	Register and grille face velocities
	Supply grille throw and spread (direction of jet)
	Pressure drop produced by terminal devices

Figure 42: Air Distribution Checklist

CHAPTER V
OVERALL SUMMARY

This guide was meant to give you some insight, and to help you discern and evaluate the design expertise of the contractor. It should help you understand what to look for to ensure the contractor has done a rigorous load calculation of the home's needs, selected heating and cooling equipment capable of meeting the heating and cooling (sensible and latent) loads, designed a network to carry the air where it is needed, and chosen the terminal devices that will properly distribute the conditioned air throughout each room. With this better understanding, hopefully you are better equipped to review the design work of professional HVAC contractors and appreciate a thorough and proper design.

Now that the HVAC system is properly designed, it needs to be followed by a quality installation. ACCA worked with several industry leaders to develop a nationally-recognized and industry-accepted Quality Installation (QI) Standard. The ACCA QI Standard identifies consensus requirements associated with quality installations, industry approved procedures for measuring or verifying the attainment of those requirements, and acceptable forms of documentation to show compliance to the requirements. These are the main elements of the ACCA Quality Installation Specification:

Design Aspects:

- Ventilation
- Building heat gain/loss load calculations
- Proper equipment capacity selection
- Geothermal heat pumps ground heat exchanger
- Matched systems

Distribution Aspects:

- Duct leakage
- Airflow balance
- Hydronic balance

Equipment Installation Aspects:

- Airflow across the indoor heat exchanger
- Water Flow through heat exchangers
- Refrigerant charge
- Electrical requirements
- On-rate for gas- fired equipment
- Combustion venting system
- System controls

System Documentation and Owner Education Aspects:

- Proper system documentation to the owner
- Owner/operator education

The ACCA Quality Installation Specification is available at no charge from ACCA's website at http://www.acca.org/quality. Free checklists are also available; these checklists offer questions to ask your contractor. If you plan to interview more than one contractor, there is a suggested method for rating each one so you have an objective way to determine the value of each contractor's proposal.

This book has taken just a little while to read, but it takes much more time and energy for the professional contractor to create a custom-designed HVAC system that will work properly. When selecting an HVAC contractor, you can use the compiled chart (Figure 43, on the next page) to ensure they have gone through this process either when the proposal is presented, or they will do it when you have accepted their proposal. There are also some helpful, free brochures at the ACCA website available for download: http://www.acca.org/tech/codes/

Now Bob has a well-designed HVAC SYSTEM that will provide a very comfortable home during winter, spring, summer and fall.

Impact	Load Calculation Issues	Equipment Selection Issues	Duct Design Issues	Terminal Selection Issues
High	A rigorous load procedure (MJ8) was used to calculate requirements	Equipment capacity is adequate for the heating and cooling loads	Manufacturer's data – blower fan performance	Register and grille velocities
High	Design Temperatures (Table 1A or local outdoor standard; MJ8 indoor default or local indoor standard)	For heating, the blower speed setting produces a safe temperature rise. For cooling, the blower speed produces adequate sensible and latent capacity[1]	Calculation of available static pressure (ASP)	Supply grille throw and spread (direction of jet)
High	Windows, Glass Doors and Large Skylights (number, size, exposure directions, U-factor and SHGC, external/internal shade)	Manufacturer's *expanded* performance data was used to select cooling equipment.	Total effective length (TEL) and room CFM values	Pressure drop produced by terminal devices
High	Ducts (location, leakage, and duct wall R-values)	Correct design parameters (CFM, outdoor temperature, EDB and EWB) were used to extract the correct equipment capacity values from the manufacturer's performance data table	Design friction rate value	
High	Ceilings under an attic (R-values, roof material, roof color)	Equipment does not exceed the recommended *(Manual S)* limit for excess capacity	Dampers installed, air flow balanced	
High			Adequate return air paths for all rooms	
Medium	Infiltration (tightness for load calc matches observed tightness)[†]	Calculate the *Manual J8* sensible heat ratio (home's SHR) to determine the target cooling CFM.	Load calculation data transfer - CFM for each room	
Medium	Ventilation	Select candidate cooling equipment based on the target airflow[2]	Duct airway flow rates and velocities	
Medium	Small Skylights	Heat pump balance point used to determine the amount of supplemental heat	Air distribution system layout (with duct sizes)	
Low	Appropriately Insulated Floors	SEER, HSPF and AFUE Ratings[3]	Register and grille velocities	
Low	Appropriately Insulated Walls and Roof-Ceilings	Interpolate manufacturer's performance data if design parameters differ by more than 5 degrees	Pressure drop produced by terminal devices	
Low	Internal Gains (normally, but some people have hobbies or needs that make this a big cooling-load item)		Location of return (regarding high or low position)	

† Infiltration is not engineered ventilation. If code requires ventilation, there shall be a ventilation load separate from the infiltration load.

1 For some applications, the same blower speed can be used for both heating and cooling, some may have homes heating and cooling system performance attributes that may allow for different heating and cooling fan speeds.

2 The important thing is that the performance of the candidate equipment has been properly evaluated (as demonstrated by this document) and found to have adequate (but not excessive) sensible and latent capacity.

3 Equipment properly selected and installed (complies with the minimum requirements of codes and the quality installation standard) has very good efficiency, though occupant comfort and total system efficiency are the relevant issues.

Figure 43: Compiled HVAC System Checklist

APPENDICES

APPENDIX A:
WHAT IF BOB'S HOUSE WERE…?
TYPICAL CHANGES THAT AFFECT LOAD CALCULATIONS

Have you ever wondered how important different factors are in calculations? What would happen if …? This Appendix explores the effects of making changes to Bob's house. Some of the changes may seem insignificant, but may prove to have a large impact on the equipment size.

Bob's house is a pretty simple ranch style house. These variances are unique to this home. Many homes have basements, or second stories, or other special architectural details to consider. A different home would react differently to these factors. This is why it is imperative to perform the heat loss and heat gain calculations for each home. The small changes we make may affect a more complex home more drastically than they affect Bob's.

Let's first consider what would happen to Bob's house if it were somewhere else. If Bob built the same house using the same materials in a different city, how would it change his heating and cooling needs? Obviously Miami and Phoenix need less heat than Chicago and Denver. But who would recognize that Denver needs less heat than Chicago – though the design temperature is lower? The answer lies in less infiltration because of the higher altitude and its lower air density.

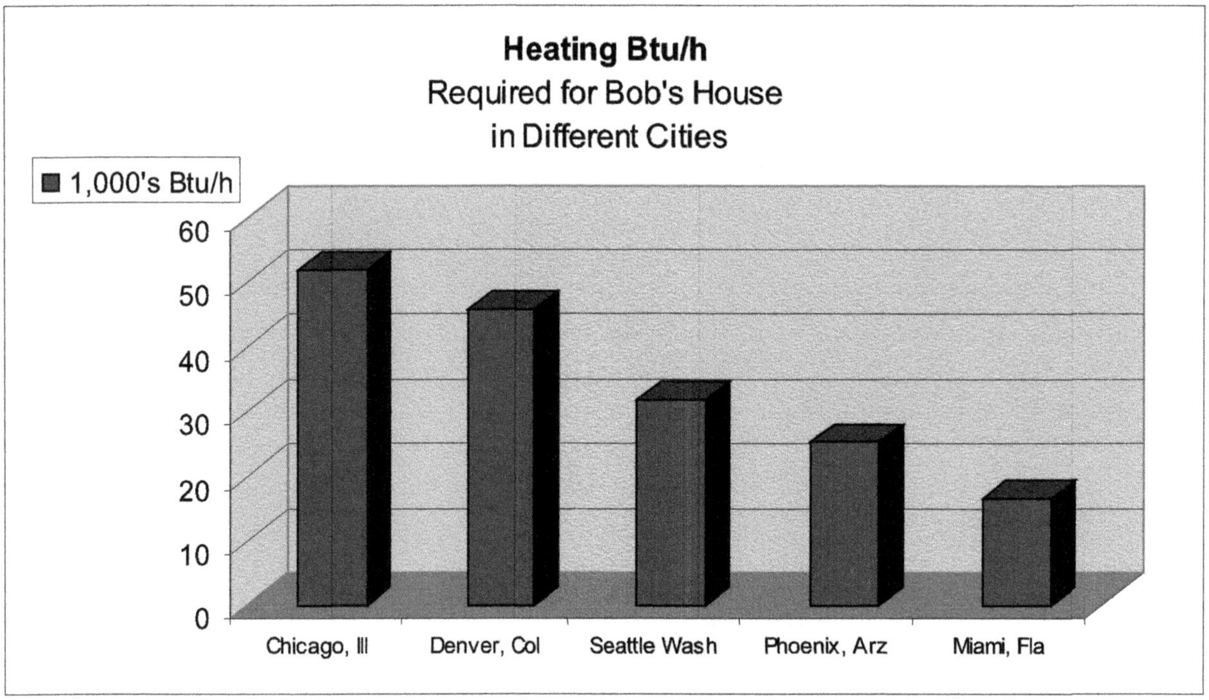

Figure A1: Comparison of Bob's Heating Needs in Different Cities

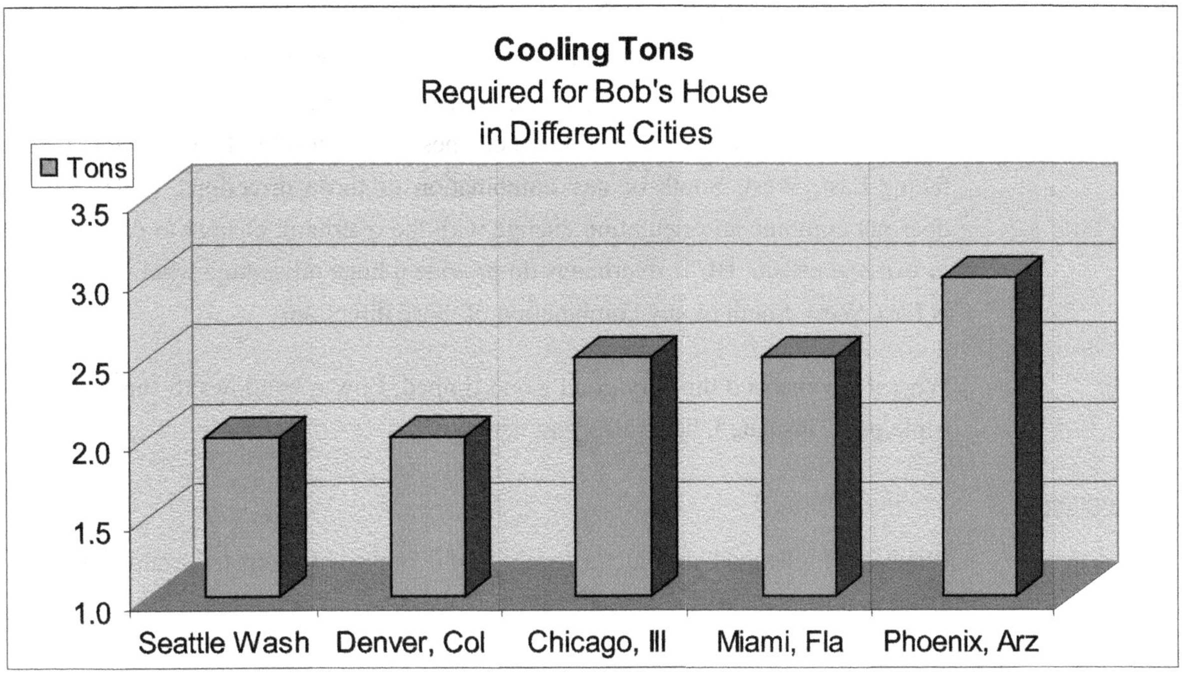

Figure A2: Comparison of Bob's Cooling Needs in Different Cities

Aside from the fact that the homes have different heat losses and gains, we must also remember the conditions under which the equipment will operate. In Phoenix, the air conditioner must provide Bob's house with 30,000 Btu/h, but the outdoor design temperature is 110°F. Normally a 30,000 Btu/h load would require about 2.5 tons, but in Phoenix, Bob needs a 3.0 ton to deliver the necessary capacity when operating under those extreme conditions.

What would happen if the builder changed one aspect of Bob's house? What effect would it have on the load calculation? The charts on the following pages depict the percentage of change in the load from changing one construction factor on Bob's house. Below are the effects of changing certain construction factors in each of the cities where Bob might live. Here are the definitions for the materials used when we consider "what if we changed?"

Wall Construction: Bob has standard "sided" frame walls with R15 insulation. What if we changed to insulated concrete forms (ICF)? This building practice uses Styrofoam blocks filled with concrete. The blocks are extremely well insulated and provide some thermal storage which can reduce the run time of the cooling and heating equipment.

Ceiling Insulation: What if we changed the insulation in Bob's attic from R30 to R44? How much will this added insulation drop the heating and cooling load? This will have a more dramatic affect in much hotter or much colder climates and a lesser effect in milder climates.

Home Orientation: What if we changed the orientation of Bob's house? Here we rotated the home to obtain the most solar heat gain through the windows as possible.

Window Overhang: What if we changed the depth of the overhang over Bob's windows to provide more shade from the sun. Bob currently has a 6 inch overhang in the example, and this increases it to 4 feet. 4 feet is a bit extreme for an overhang, but with awnings and decks this exists. The overhang becomes vital when the home's windows are facing East, West, South or any combination of those directions. The example does not combine an orientation change with the overhang change to create a false or extreme effect. BUT, overhangs do provide a huge advantage where orientation is East, West, South or any combination of those directions.

Windows: What if we changed the very good wood framed, Low-e windows to super-efficient triple-pane, insulated, heat reflecting window?

Duct Location:

Attic: What if we changed the duct's location and moved it from the ventilated crawl space to a vented attic?

Conditioned Space: What if we changed the duct's location and moved the supply and return ducts from the ventilated crawl space inside Bob's House, the conditioned space? In homes with multiple levels, this could be the area in between the basement and the first floor, or the space between the first and second floor. In a single level home the duct system could be located in soffits throughout the home. This change requires some coordination between the generals and HVAC contractor, however it can be done.

Duct Sealing and Insulation: What if we changed the duct leakage and duct insulation value? Bob's house has partially sealed ducts and standard R-6 insulation. What would happen if we sealed the ducts to a higher level and increased the insulation value to the R-8 level required by the 2015 Residential Code (International Code Council)?

CHICAGO

In Figure A3 we see that the cooling load (as with most of the scenarios) benefited most from locating the ducts in the conditioned spaces, between floors or in soffits. Upgrading the windows and sealing the duct system also greatly reduced the cooling load.

However, if Bob's house was rotated to face Northeast, and his ducts were relocated into the attic, his load would have increased by 11%, and his air conditioner size would have gone up.

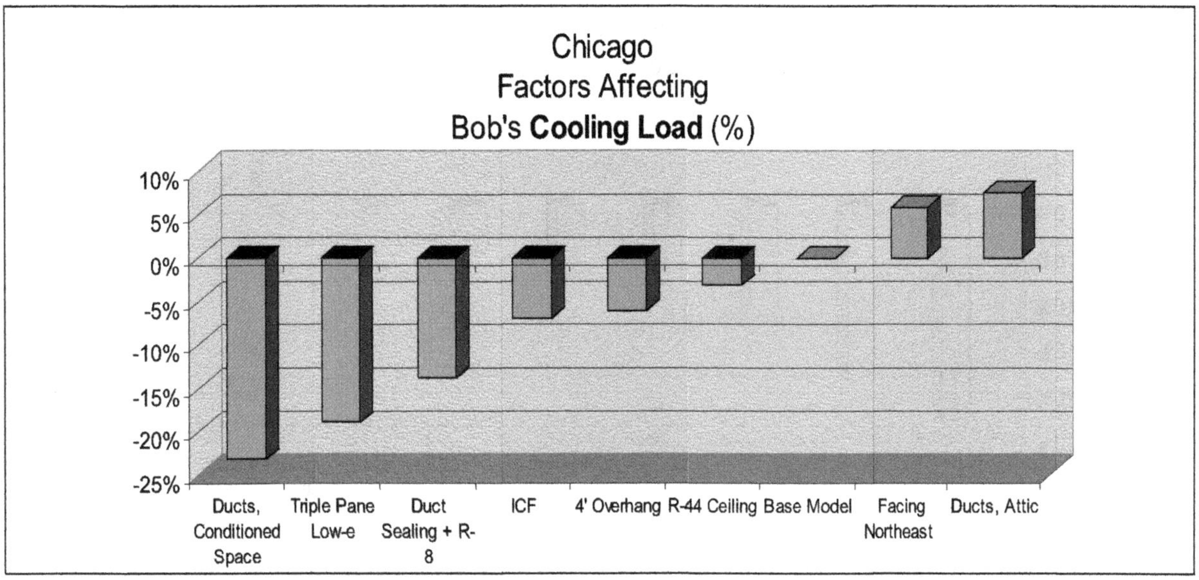

Figure A3 – Factors Affecting Bob's Cooling Load – In Chicago

The heavier and better insulated ICF, duct sealing, and duct insulation would greatly help Bob save on first cost for smaller HVAC equipment, and he might enjoy more even temperatures, at no real added expense.

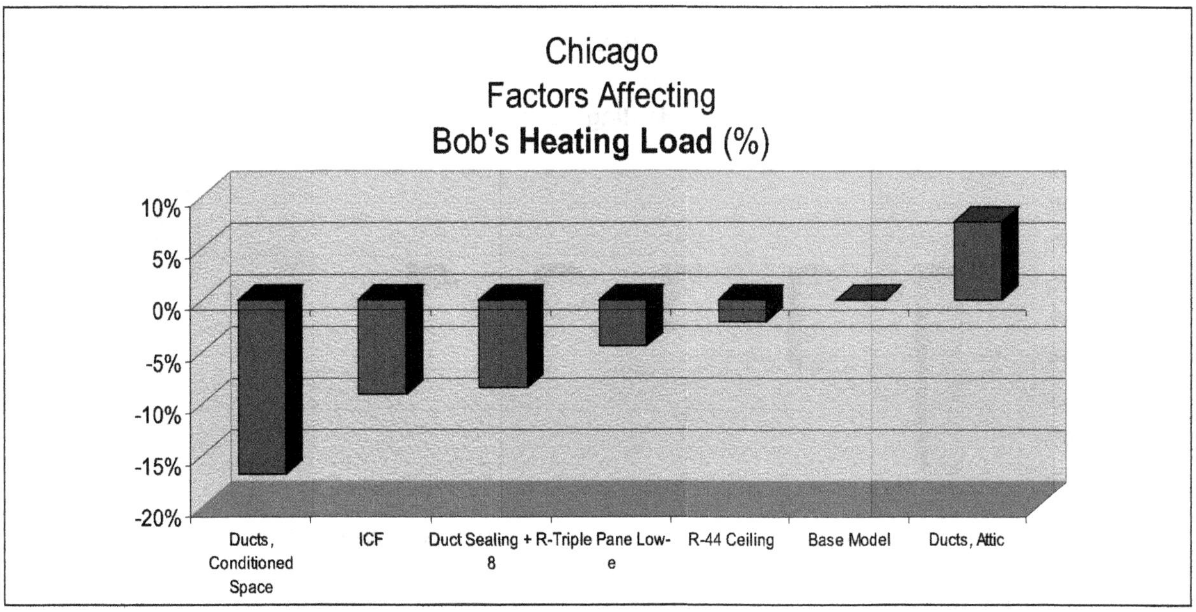

Figure A4 – Factors Affecting Bob's Heating Load – In Chicago

MIAMI (Hot and Humid)

In the hot and humid Miami area, Bob benefited most (besides ducts in the conditioned space) from good windows, duct sealing and better duct insulation.

Here too, if Bob's house faced northeast and his ducts were in the hot attic, he would need to purchase a larger unit and spend more to cool his house.

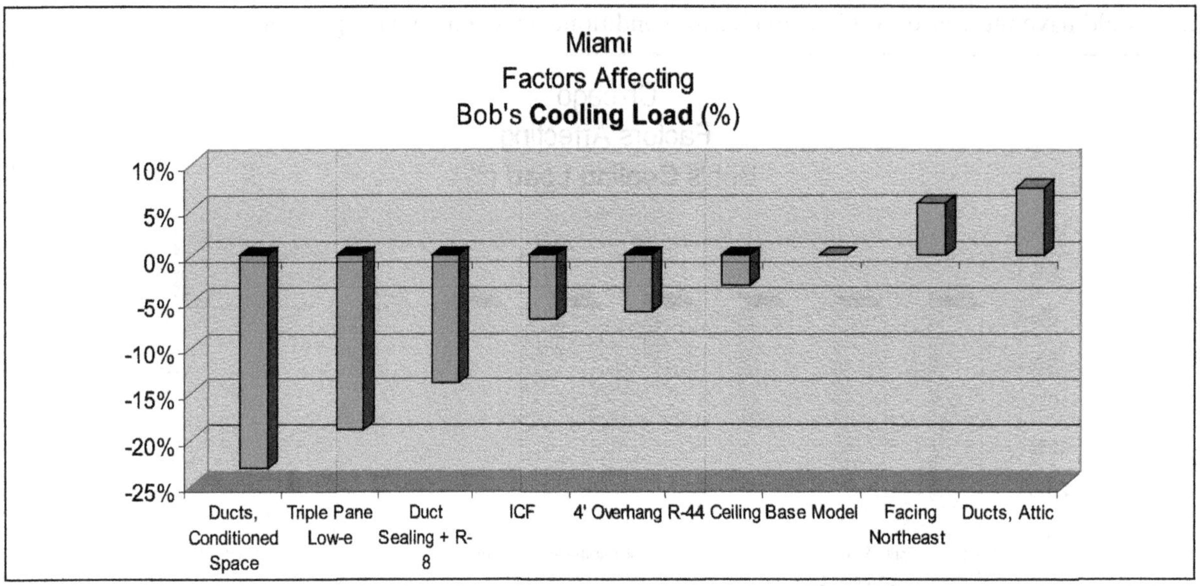

Figure A5 – Factors Affecting Bob's Cooling Load – In Miami

Heating? In Miami? Well, for those days when the temperature drops to 60°F, here are the big contributors to the load. Ducts in the attic or in conditioned space (relocated from the base example of being in the crawlspace) affect Bob's equipment costs the most. If Bob is using an electric heat pump or electric auxiliary heat, then his operating costs will go up as well if the ducts are in the attic.

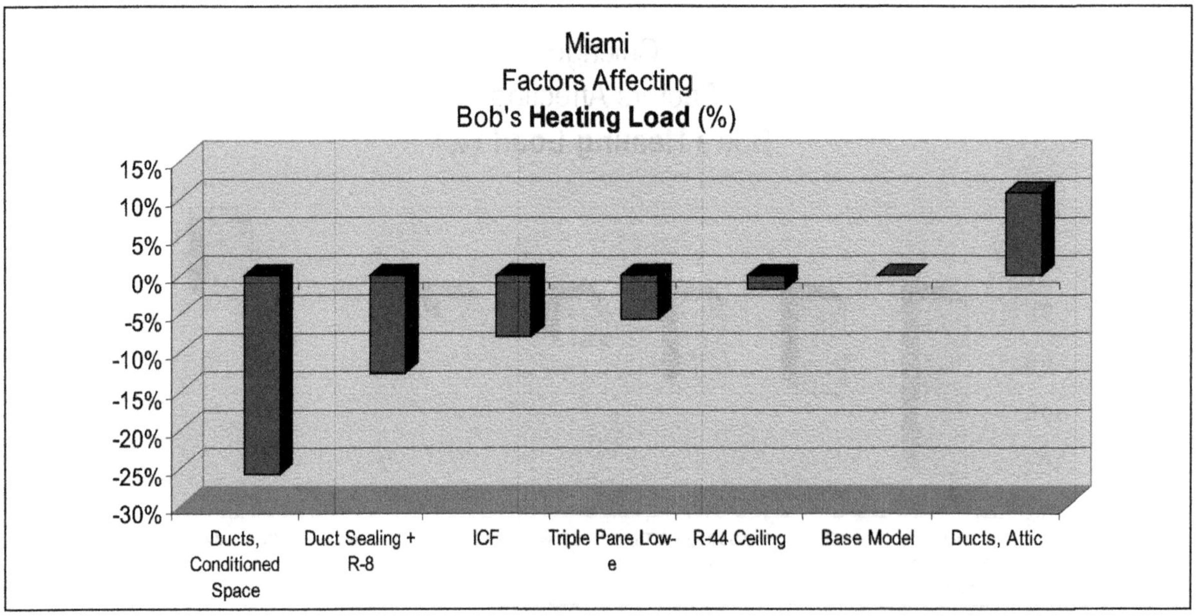

Figure A6 – Factors Affecting Bob's Heating Load – In Miami

PHOENIX (Hot and Dry)

This is one hot town, and one of the few places where installing better windows has the same effect as relocating the ducts to the conditioned space. Here the big temperature difference also made the ICF walls a good choice. Maybe those Navaho Indians knew something when they built with adobe. Again facing Northeast with ducts in the oven, I mean attic, would cost Bob some money.

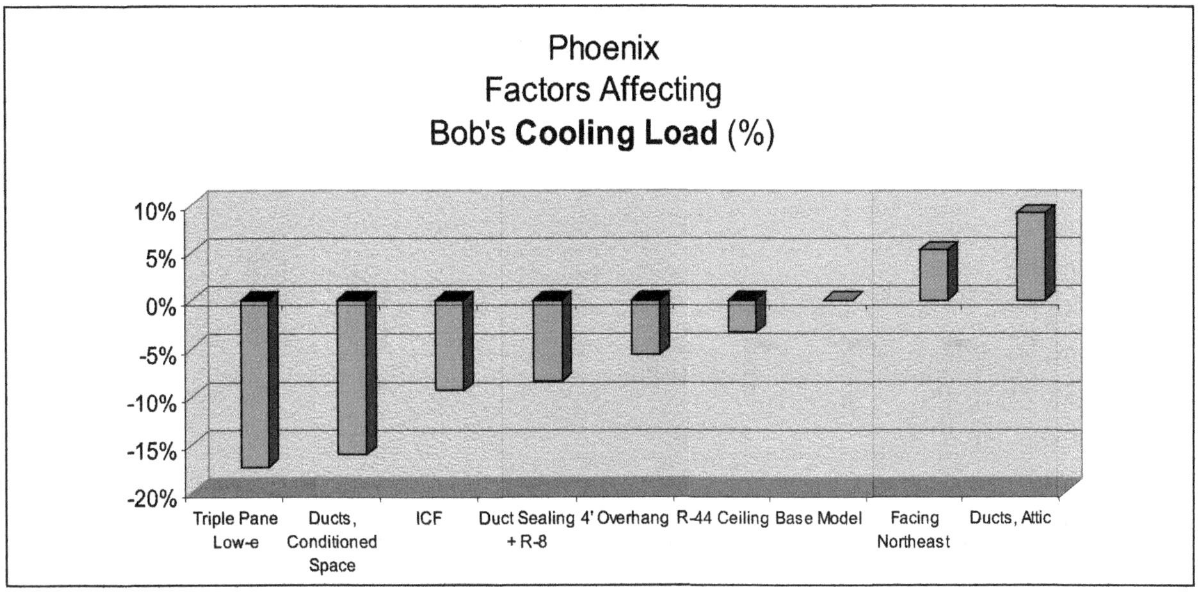

Figure A7 – Factors Affecting Bob's Cooling Load – In Phoenix

Heating in Phoenix is like heating in Miami, except the dry air can make 60°F feel very uncomfortable without a jacket. Moving the ducts inside, or sealing them well, insulating them better, and building houses like the Navaho reduces heating needs.

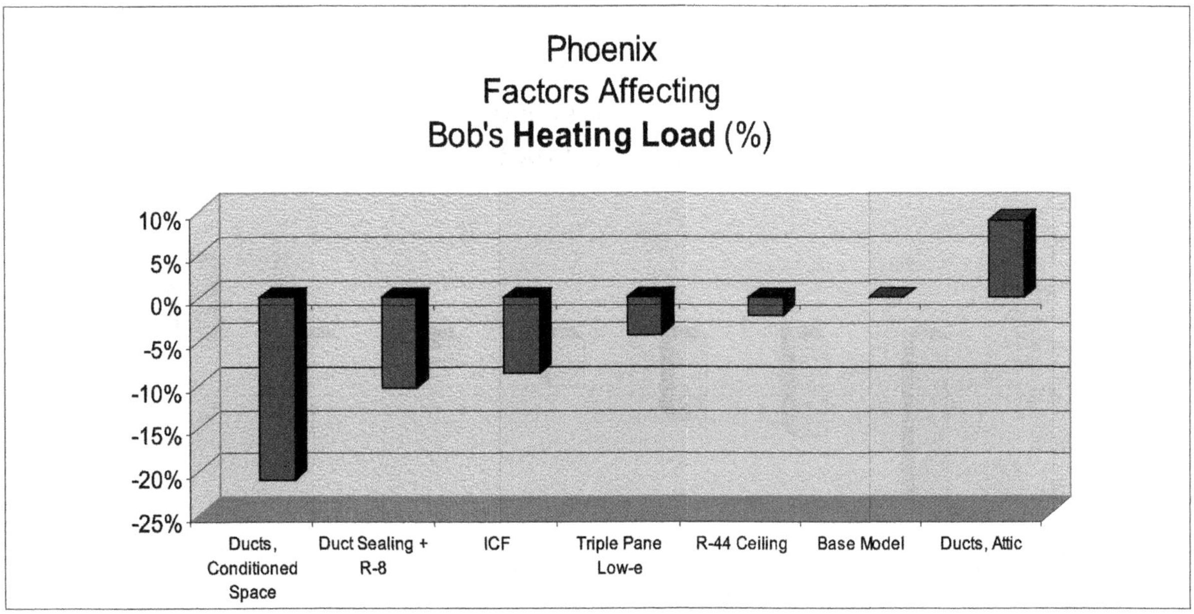

Figure A8 – Factors Affecting Bob's Heating Load – In Phoenix

DENVER (Cold and Dry)

Cooling in Denver is like heating in Miami. For the COOLING mode, the better windows made the biggest difference. The heavier ICF walls and relocating the ducts to the conditioned areas of the home would greatly reduce Bob's cooling needs. Contractors sizing equipment and determining airflow must remember to allow for the Altitude Correction Factor. This factor plays a significant role and can alter the needed size of the equipment.

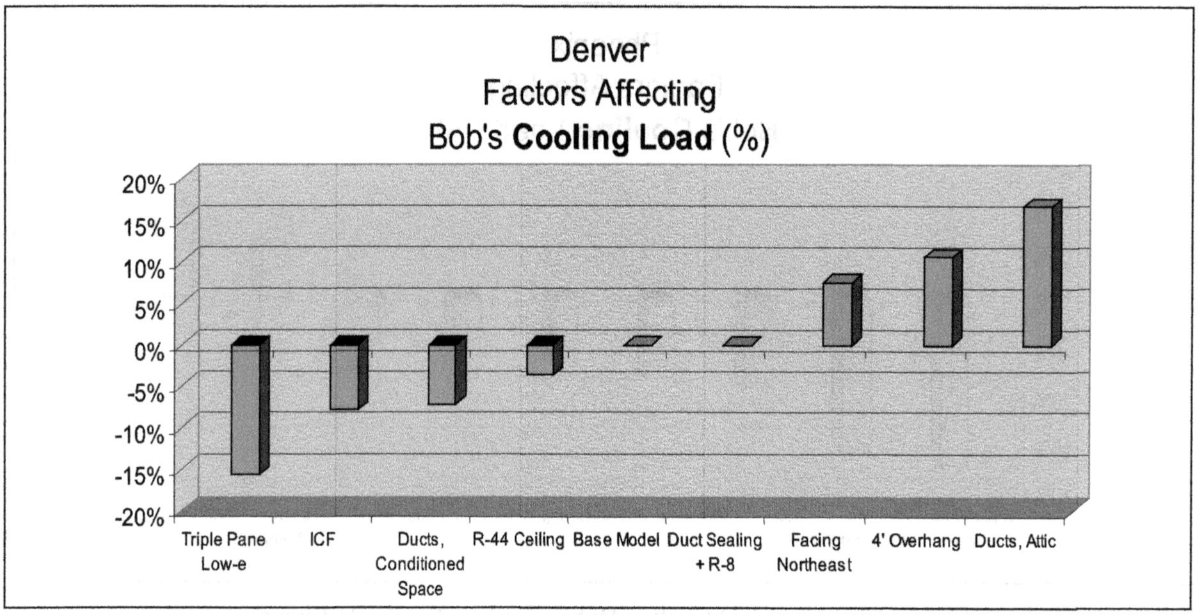

Figure A9 – Factors Affecting Bob's Cooling Load – In Denver

Heating Bob's house brings the ducts in conditioned space back to the forefront. There is a great deal of heat that is lost and gained through the duct system. When the ducts are sealed and well insulated, or ideally, located in conditioned space, then more heat goes to making Bob comfortable and less is spent on heating or cooling in Denver, or Miami, or Chicago…

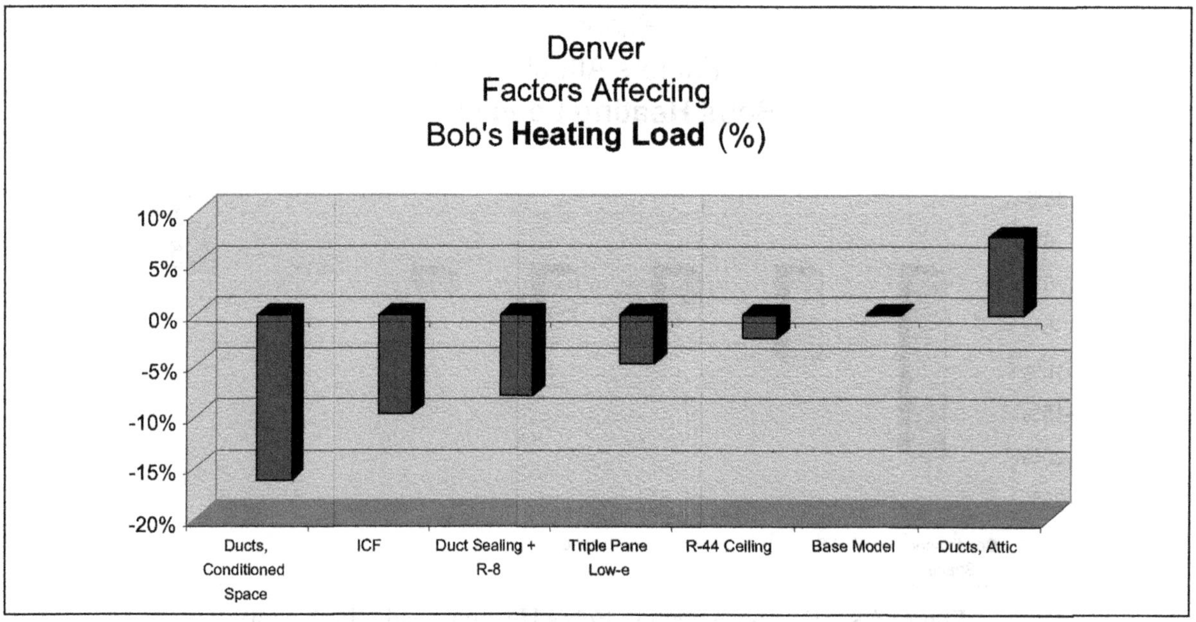

Figure A10 – Factors Affecting Bob's Heating Load – In Denver

SEATTLE (Mild)

Seattle has a very mild climate. You may have noticed that it needs the least amount of cooling and was in the middle of the pack for heating. However, better windows and good ducts would save Bob money on equipment and operating costs.

Even in Seattle, if Bob's house was facing Northeast and his ducts were in the attic, he would need bigger equipment and would pay higher utility costs to keep his house at a comfortable temperature.

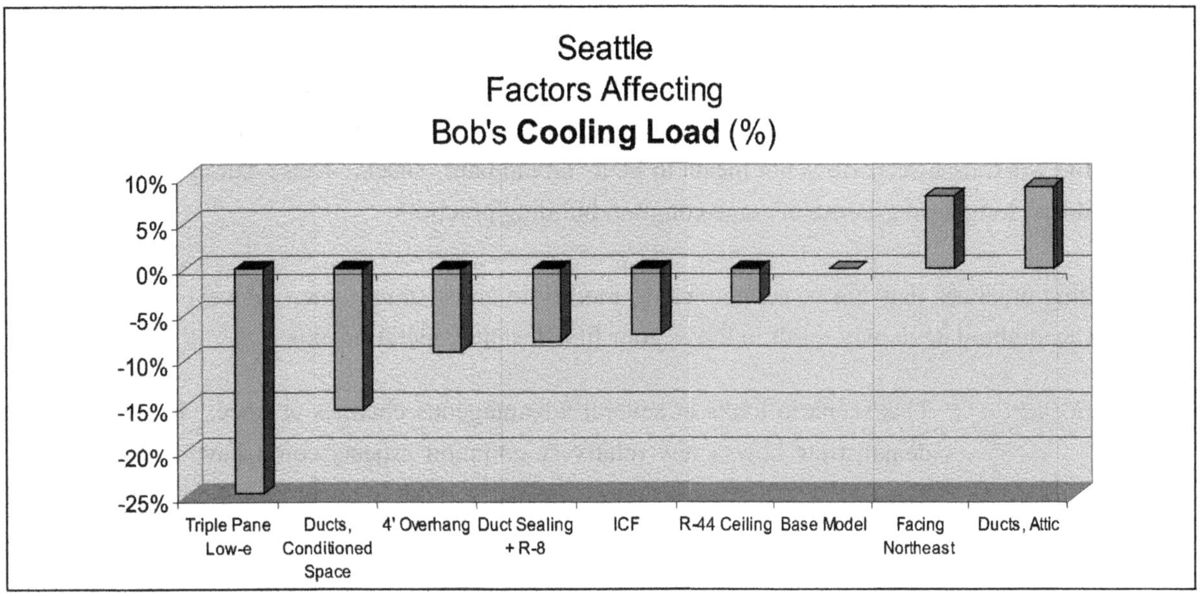

Figure A11 – Factors Affecting Bob's Cooling Load – In Seattle

Duct location, duct sealing, and solid ICF walls can make a big difference in Bob's bottom line.

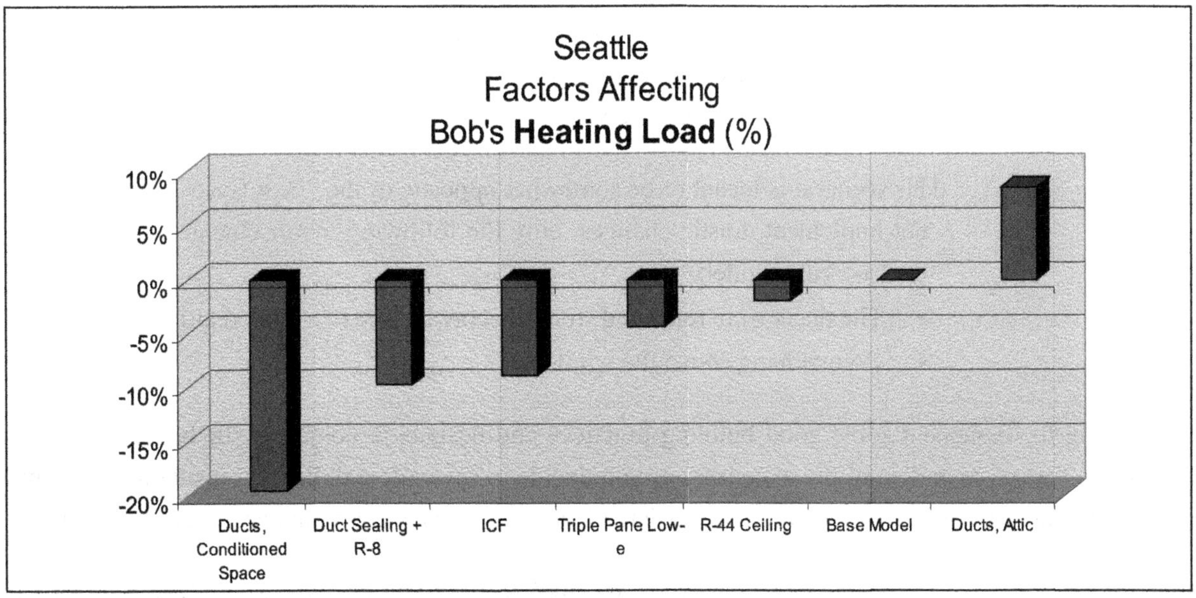

Figure A12 – Factors Affecting Bob's Heating Load – In Seattle

It Was The Best Of Times, It Was The Worst Of Times…

What if we made a best case and a worst case scenario? What if we took some of the better construction types and combined them? What would happen if some of the worse construction types were used?

We do not want to extravagantly stack the deck to create a habitable thermos bottle, but there are a few good building practices which would be relatively inexpensive. Some changes can be made to existing homes to improve their comfort and reduce their energy use, e.g., it is a relatively minor added expense to add more insulation to an attic, but changing from standard stick-built to insulated concrete forms (ICF) is a very large expense. If you were building a home and had to make construction material choices you may choose ICF construction for its numerous building qualities. There are also regional building practices which will help, like deeper window overhangs will help in areas with lots of sunshine. Likewise, the worst case scenario is not meant to be a "thread bare" shack. These examples merely point out the "hidden" long term expense of some common building practices.

Some building practices may not be economical to modify in an existing home. However, if a change is possible to accommodate, it may result in lower your first cost and operating expenses.

<u>Best Case Scenario</u>: This scenario looks at several advantageous changes at once. It was meant to demonstrate how a few relatively common aspects could have a large impact. The Best Case Scenario takes the "base model" home and modifies it with the following changes:

- The insulation in attic is boosted from R-30 to R-44,
- The double-pane Low-e windows are replaced with triple-pane windows which have a reflective surface,
- The overhang was extended from 6 inches to 3 feet (we found that a 2.5 foot overhang with a gutter was more common than a 4 foot overhang),
- The ducts were relocated from the crawlspace to the inside of Bob's house, in conditioned space. This is the only change that might be difficult to affect simply and inexpensively.

<u>Worst Case Scenario</u>: This scenario is meant to be somewhat opposite of the "Best Case". In an effort to not implement drastic changes, only the following minor changes were made to the base case model:

- The ducts were relocated from the crawl space to a vented attic, and
- No overhang above the windows.

By deciding to implement some good building practices and designs, a contractor or home owner can save on the first costs and on utility expenses. If you decide to save a few dollars on insulation, you may spend it on more expensive equipment, the associated installation costs, maintenance, and utility costs.

Here are the results of the equipment sizes when we applied the best case and worst case scenarios to Bob's base model house:

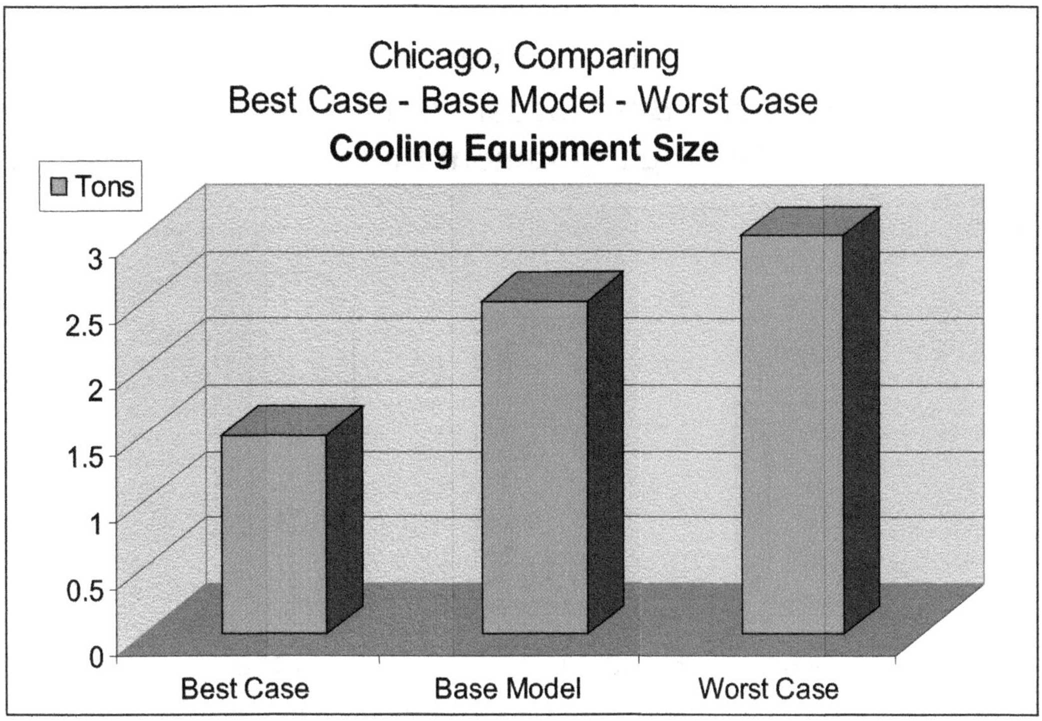

Figure A13 – Chicago Comparison

In Chicago, Bob's air conditioner drop by one ton in the best case scenario. A smaller unit has a lower first cost and a lower operating cost.

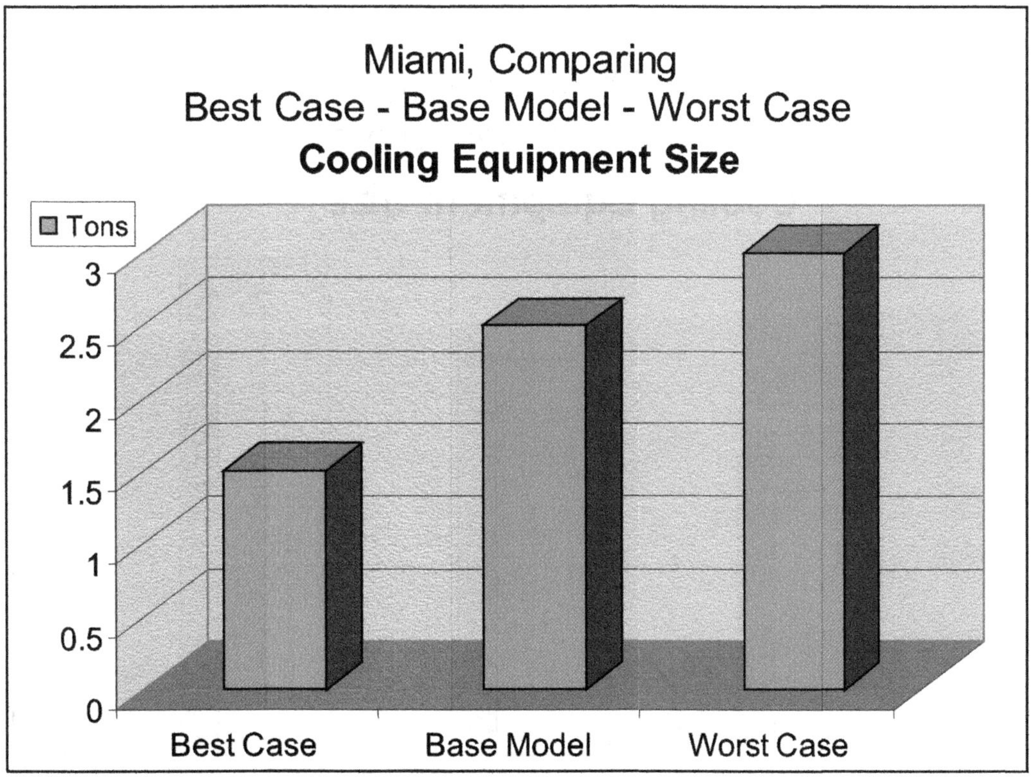

Figure A14 – Miami Comparison

In Miami, the results are very similar, with the ducts in the attic the air conditioner size grew from a 2.5 ton to a 3.0 ton.

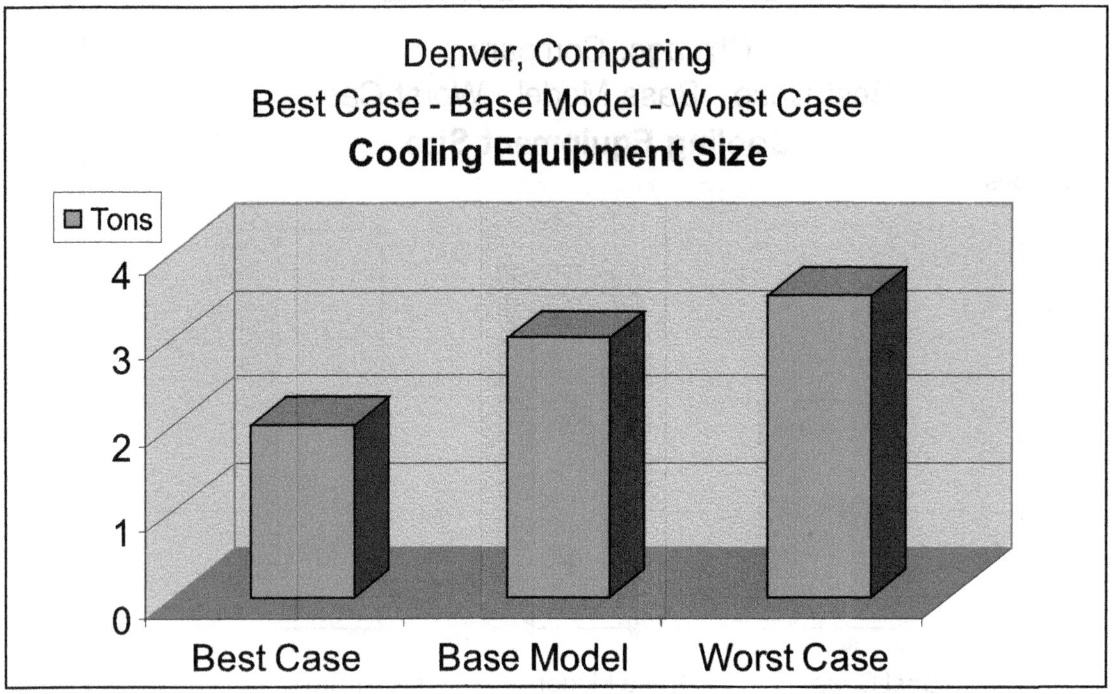

Figure A15 – Denver Comparison

In Denver, cooling equipment size must be altered to account for the altitude. There is still one ton of capacity that can be saved by using the best case scenario. The same is true for Phoenix below.

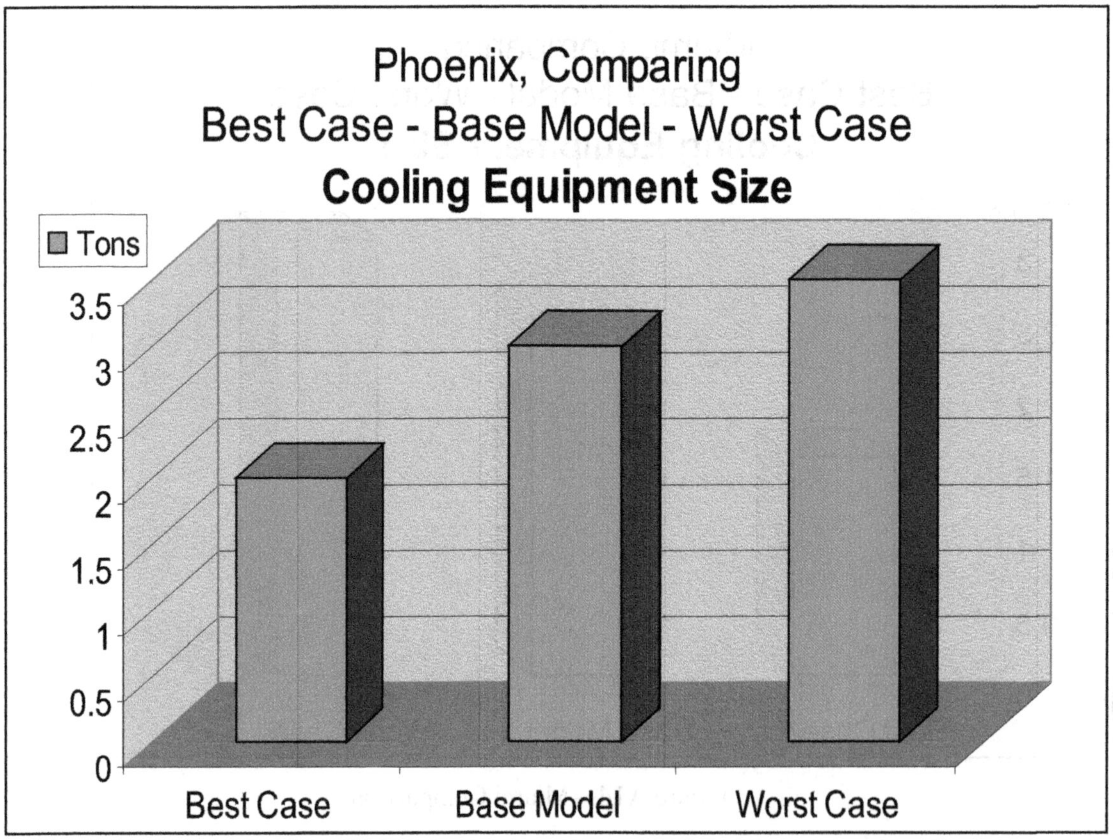

Figure A16 – Phoenix Comparison

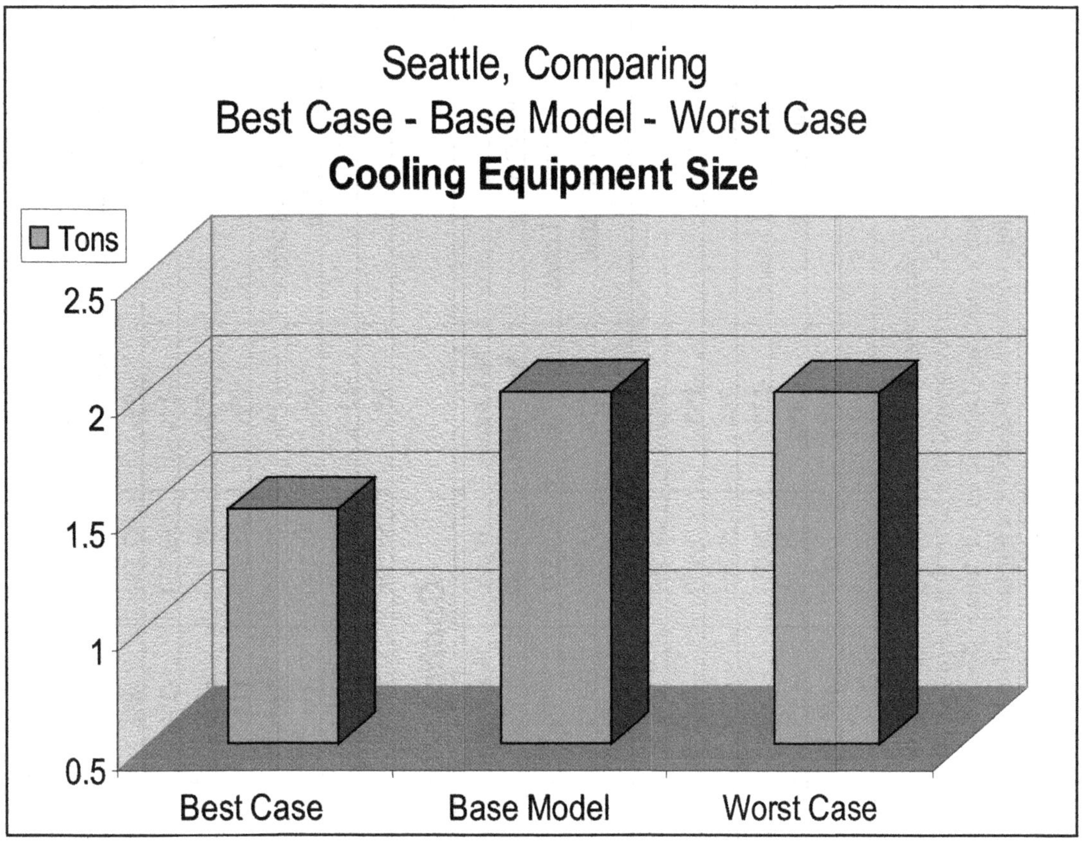

Figure A17 – Seattle Comparison

In Seattle's very mild climate, the difference between the best and worst case is only 0.5 ton. However, the first costs and operating costs, over time will add up. If everyone in Seattle applied some simple modifications to their homes and changed the size of their air conditioner, the city could drastically reduce its carbon foot print.

The table on the next page provides the actual numbers that support the charts used above.

Heating: What if Bob's House were: ? ? ?

	Chicago, Il		Miami, Fl		Phoenix, Az		Denver, Co		Seattle, Wa	
	Cold & Humid		Hot & Humid		Hot & Dry		Cold & Dry		Mild	
	% Change	Btu/h	**% Change**	Btu/h	**% Change**	Btu/h	**% Change**	Btu/h	**% Change**	Btu/h
Bob's "Base" home[1]	**Base Model**	51,838	**32%**	16,526	**49%**	25,229	**88%**	45,852	**61%**	31,768
Insulated Concrete Forms[2]	**-9%**	47,101	**-8%**	15,192	**-9%**	23,027	**-10%**	41,382	**-9%**	28,966
R44 Insulation in Attic	**-2%**	50,754	**-2%**	16,221	**-2%**	24,725	**-2%**	44,829	**2%**	31,127
Triple Pane Low-e[2]	**-5%**	49,500	**-6%**	15,563	**-4%**	24,142	**-5%**	43,646	**-4%**	30,386
Ducts, Attic[2]	**8%**	55,776	**11%**	18,277	**9%**	27,499	**8%**	49,367	**8%**	34,459
Ducts, Conditioned Space[2]	**-17%**	43,112	**-26%**	12,248	**-21%**	19,928	**-16%**	38,412	**-19%**	25,577
Duct Sealing & R-8	**-8%**	47,436	**-13%**	14,406	**-10%**	22,597	**-8%**	42,224	**-10%**	28,667
Best Scenario[3]	**-23%**	39,691	**-32%**	11,284	**-27%**	18,337	**-23%**	35,183	**-26%**	23,553
Worst Scenario[4]	**8%**	55,776	**11%**	18,277	**9%**	27,499	**8%**	49,367	**8%**	34,459

Cooling: What if Bob's House were: ? ? ?

	% Change	Btu/h	**% Change**	Btu/h	**% Change**	Btu/h	**% Change**	Btu/h	**% Change**	Btu/h
Bob's "Base" home[1]	**Base Model**	26,107	**77%**	28,039	**70%**	30,732	**103%**	20,859	**111%**	19,403
Insulated Concrete Forms[2]	**-7%**	24,331	**-7%**	25,989	**-9%**	27,821	**-8%**	19,213	**-7%**	17,991
R44 Insulation in Attic	**-3%**	25,324	**-3%**	27,193	**-3%**	29,689	**-4%**	20,106	**-4%**	18,704
Facing Northeast[2]	**6%**	27,668	**6%**	29,844	**5%**	32,335	**8%**	22,445	**8%**	20,945
4' Overhang[2]	**-6%**	24,564	**-4%**	26,948	**-6%**	28,966	**-11%**	23,066	**-9%**	17,639
Triple Pane Low-e[2]	**-19%**	21,189	**-19%**	22,698	**-17%**	25,372	**-16%**	17,592	**-25%**	14,606
Ducts, Attic[2]	**7%**	28,105	**7%**	30,062	**9%**	33,522	**17%**	24,348	**9%**	21,145
Ducts, Conditioned Space[2]	**-23%**	20,087	**-25%**	21,067	**-16%**	25,791	**-7%**	19,356	**-15%**	16,413
Duct Sealing & R-8	**-14%**	22,523	**-15%**	23,788	**-8%**	28,127	**-8%**	19,284	**-8%**	17,848
Best Scenario[3]	**-46%**	14,017	**-44%**	15,583	**-38%**	19,124	**-34%**	13,668	**-46%**	10,497
Worst Scenario[4]	**7%**	28,105	**8%**	30,210	**9%**	33,522	**18%**	24,705	**9%**	21,145

1. Bob's "Base" home compares the same home in five different cities.
2. Bob's "Base" home is compared to a home that has just one different orientation or building construction type. For example, Insulated Concrete Forms (ICF) compares Bob's stick built home to an ICF home. The modification for each example is the only modification made. There are no cumulative additions (i.e., the orientation, windows and ducts were not factored in all at once).
3. Best Scenario is the "base" home with the following changes: R-44 insulation in attic, triple-pane windows, three foot overhang, and ducts in conditioned space.
4. Worst Scenario is the "base" home with the following changes: ducts relocated to a vented attic and no overhang over the windows.
5. These variances are unique to this home. A different home would react differently to the listed factors. This is why it is an imperative to perform the heat loss / gain calculations for each home.

APPENDIX B

COMMENTS ABOUT OTHER EQUIPMENT TYPES

The concepts and procedures used for a gas furnace and electric cooling equipment apply to other common heating and cooling systems. In some cases there may be a minor difference or a small amount of additional work.

- The concept of input and output capacity and adequate blower CFM for cooling applies to gas, oil, propane and electric furnaces.
- The procedure for electric cooling applies to air source heat pumps in the cooling mode see *Manual S*, §6 :
    - o First size the equipment for cooling (exactly as for an air conditioner).
    - o Draw a balance point diagram (see Figure B-1).
    - o Use the balance point diagram to determine the supplemental heat requirement
- Determine the additional (electric or fossil fuel) heat requirement (see *Manual S*, §8.)

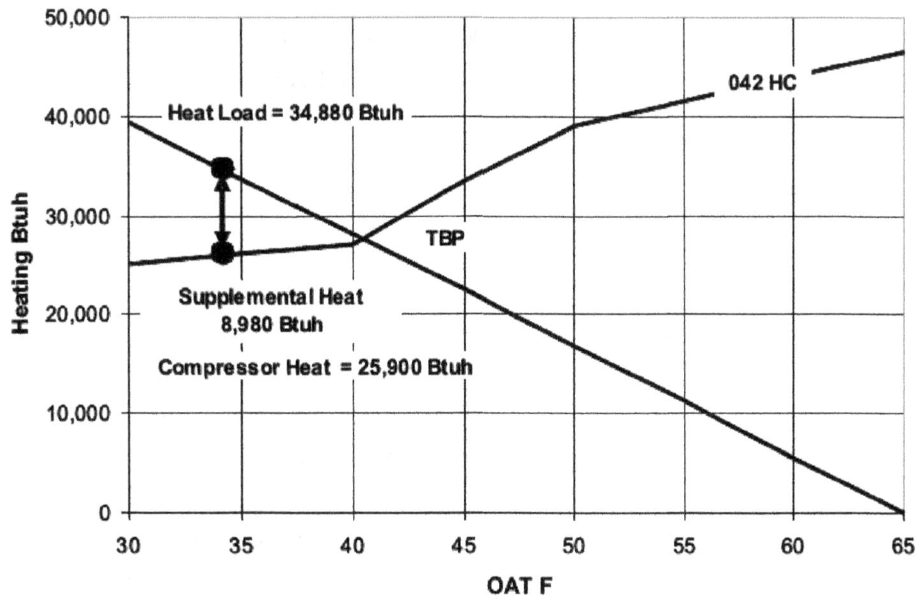

Figure B1: Sample Balance Point Diagram

- The procedure for air source heat pumps applies to water source heat pumps (see Manual S, §9:
    - o First, size the equipment for cooling (exactly as for an air conditioner, substitute entering water temperature for outdoor air temperature).
    - o Draw a balance point diagram.
    - o Use the balance point diagram to determine the supplemental heat requirement.

    Determine the additional (electric or fossil fuel) heat requirement.
- The furnace procedure and the air source heat pump procedure are used for a dual fuel system; evaluating the economic balance point adds a little more work (see *Manual S*, §8).

- Do not oversize an air source or water source heat pump to get a lower balance point (See Appendix F: i.e., do not exceed *Manual S* limits on excess cooling capacity). Oversized equipment increases costs associated with cycling the equipment on and off repeatedly in heating and cooling seasons.
- For heat pumps, supplemental heat and auxiliary heat (total strip heat) are two different things. The amount of supplemental heat (i.e., heat activated by the second stage of the thermostat) is determined by the balance point diagram (rounded to the closest standard kilowatt size). Total strip heat is typically the kilowatt (KW) equivalent of the design heating load, which may be much larger than the supplemental KW. During an emergency, the controls should activate the supplemental heat, plus an increment of auxiliary heat that is not available during normal operation.

Incorrect guidance is floating around the HVAC and utility industries. Some of this finds its way into manufacturer's literature, local codes and utility standards (often to the benefit of some vested interest and to the detriment of the homeowner). One prime example of incorrect guidance is, oversizing multi-speed and variable-speed or variable-refrigerant-flow equipment. Equipment sizing is not changed for any of the following types of equipment if a variable operation type of equipment is selected:

- Geothermal (water source) Heat Pump: **Never size an air source or water source heat pump for the heating load**. These heat pumps transfer heat to a body of water. The water may be enclosed in a piped loop, or come from a lake, or well.

- Ductless Heat Pumps and Air Conditioners: The same sizing rules apply. Depending on the manufacturer, the equipment may have a variable-speed indoor and outdoor blower motor, and a variable-speed compressor. These features allow the unit to address partial load conditions very effectively. Some of the heat pumps may also need less, or no, auxiliary heat because they are able to adjust their balance point by adjusting their head pressure. Reviewing manufacturer's data becomes very important when adding auxiliary heat. However, the equipment is still sized based on the peak load conditions, and the supplemental heat and auxiliary heat are still based on the balance point.

- High Velocity Fan Coils: These unit's duct systems emit a strong jet of air from small, 2" diameter diffusers. The air leaves at a very high velocity. The shape of the duct will determine when the air will reach terminal velocity. The air will reach terminal velocity in about 1.5 to 2 feet, if the duct is installed correctly. Contractors should take great care to locate the supply diffuser so the occupants are not bothered by drafts. The high velocity fan coils also use a different method for duct system sizing. Consult the OEM's installation instructions to properly size the duct system.

- Evaporative Coolers ("Swamp Coolers"): These units use the heat transfer of evaporation to condition air for the home. The cooler uses a water source to saturate a media, a fan draws across the media, the water cools and humidifies the air, and the conditioned air is distributed into the home. The source of the air is 100% outdoor air, no return ducts. An air path to the outside, usually a window, must be opened to provide circulation. The cooling performance of a single-stage evaporative cooler is determined primarily by the temperature and relative humidity of the incoming air. Neither ACCA nor *Manual S* provides guidance for selecting an evaporative cooler.

APPENDIX C
DUCT SYSTEM ISSUES AND CHALLENGES

There are a number of problems that are caused by undersized ducts, added accessories, and poor air circulation in the home. Each causes its own set of problems.

Dampers: Duct systems must have balancing dampers (per ANSI/ACCA *Manual D*). Balancing dampers provide a way for the HVAC contractor to adjust slightly oversized ducts to the desired value. Some adjustments can be done at supply registers, but if they are seriously throttled, they will produce undesirable noise. Oversized ducts can cause problems if there are no balancing dampers[39].

Undersized Duct Systems: These cause indoor blower assemblies to produce excess air pressure. Excess pressure can cause leaks. Leaky, restricted ducts will cause:

- Clean, conditioned air to leak out of the supply ducts, and unconditioned, dusty, polluted air to be drawn into the return ducts. The excess pressure will also affect furnaces or air handlers as air is pushed from and drawn into the cabinets.
- An increase in temperature rise at the furnace heat exchanger during heating, which could possibly damage the heat exchanger.
- Incorrect refrigerant pressures and temperatures, possibly tripping a safety control or damaging the equipment.
- Inadequate cooling capacity. Remember, manufacturer's expanded performance data is only correct for the blower CFM listed on the data sheet. If the CFM through the cooling coil changes, the total, sensible, and latent capacities of the equipment change.
- Frosting or icing on the cooling coil.
- Equipment efficiency is diminished; more energy is used to provide less comfort.

Air-Side Devices: Another culprit is air-side devices that may have been ignored during the duct sizing procedure, or retrofitted post-facto. After-market, high-efficiency filters can be very restrictive and magnify existing problems. These filters can cause the same problems as restricted ducts.

Return Air Paths: Duct systems must provide circulation paths into and away from each room in order to prevent problems. Rooms with poor circulation can create a number of problems, such as:

- Negative pressure which causes excess carbon monoxide production in fossil fuel appliances and may draw lethal levels of those dangerous fumes into the home.
- Negative pressure that increases infiltration, thus introducing pollutants, allergens, and contaminants from the outdoors or unconditioned spaces (i.e., exhaust fumes from the garage).

---

[39] Note: It is generally not practical to size every section of duct run so that it is neither undersized nor oversized. Even an accurate *Manual D* design induces "some minor oversizing" in some circumstances. This is why balancing dampers are so important. It is a lot easier to adjust the balancing damper in an oversized duct than it is to replace the duct with a larger version, add a parallel "helper duct", or remove the blower and replace it with one that has more power.

- Furnace safety devices sense back drafting and stop equipment operation.

Measured Furnace Temperature Rise: Heat exchanger temperature rise decreases as airflow increases[40]. It is one thing to verify that the rise is acceptable during the design process, but quite another thing to obtain that rise after the duct system has been installed. Incorrect fan speed settings, or oversized (un-dampered) ducts can cause heating season condensation in the heat exchanger. The condensation forms acid which will damage the equipment.

A majority (perhaps most) of the duct systems in American homes are undersized. The main reasons for incorrect duct sizing are:

- Honest mistakes are made in calculating the friction rate. A good HVAC contractor must spend considerable time measuring and computing. Some common mistakes are:
  - used incorrect input for the blower's external static pressure
  - used incorrect pressure drop across airside devices
  - used incorrect total effective length of the air circulation path
  - used a guess or rule-of-thumb method.
- The friction rate used for duct sizing is incorrect for the type of duct material. For example, flexible wire-helix duct produces much more resistance than sheet metal duct. (The ACCA duct slide rule provides friction rate scales for metal duct, duct board, duct liner and flexible duct.)
- An inappropriate and inadequate "rule of thumb" was used for duct slide rule calculations. The guidance to *"use a 0.10 friction rate to size any and all duct runs"* offered by some equipment manufacturers, wholesaler trainers and vocational school teachers, and which appears on many duct sizing slide rules, is a lot of nonsense. This friction rate is occasionally correct by pure luck.
- A totally irrelevant method[41] used to determine room airflow rates. This concept is so absurd, it makes the previously noted nonsense look like sound engineering. Room CFM is determined by thermodynamic law and depends exclusively on room load, the temperature of the air supplied to the room, the room temperature and altitude. (See pages 13-15 for the process used to calculate the target airflow for Bob's house.)
- A guess is made on the load calculation, so a guess is made about the volume of air for a room.
- The effective length of the fittings is ignored or underestimated. One inefficient fitting can add 30 to 100 feet or more to the effective length of a duct run. Therefore, a system with efficient fittings can be 100 to 200 feet shorter than the same system with a few inefficient fittings.
- The installer used a field-fabricated fitting with an abrupt squeeze on airflow can increase the total effective length of a duct run by over 100 feet.
- Another cause of excess pressure is internal insulation. Let's assume that a 12 x 8 sheet metal duct is needed to carry about 500 CFM. A 12 x 8 sheet metal that uses an internal one inch insulating liner must have one inch added to each side of the duct, thus becoming a 14 x 10 duct. Also, the internal liner has a different friction rate than smooth sheet metal. Designers must be careful to resize the duct for the duct liner – a 12 x 8 duct can grow to 16 x 10 when all factors are considered.

---

[40] Furnace or air conditioner safety devices detect out-of-range temperature or pressure and stop equipment operation.

[41] This is the "room CFM equals room volume or floor area multiplied by some magic number" method.

APPENDIX D
ILLUSTRATIVE EXAMPLE OF INFORMATION PRESENTED TO CODE OFFICIALS

This Appendix provides examples of information that code officials may request from HVAC contractors. Some of the information may exist in the local building code amendments.

The information presented to a code official could include the following:

1. Design temperatures for winter and summer.
2. Design Entering Wet Bulb temperature
3. Total calculated heat loss
4. Calculated heat gain
    a. Total heat gain
    b. Sensible heat gain
    c. Latent heat gain
5. The buildings Sensible Heat Ratio (for targeting desired airflow)
6. Heating Equipment
    a. Manufacturer
    b. Model
    c. Output Capacity, (Must not exceed 140% of #3[42])
    d. Required Temperature Ranges
7. Blower Data
    a. Airflow
    b. External Static Pressure (to achieve airflow)
8. Cooling Equipment ***(From Manufacturer's Expanded Performance Data)***
    a. Manufacturer
    b. Model
    c. Total Capacity (rated at design airflow; must not exceed 115% of #4[43])
    d. Sensible Capacity
    e. Latent Capacity
9. Heat Pump Auxiliary or Supplemental Heat
    a. Heater Type
    b. Heater Capacity

[42] Heat pumps would use the heating capacity at the design outdoor temperature.

[43] In predominately heating climates, a heat pump may be sized at 125% of the cooling load.

Furnace and Air Conditioner Example

Residential Plans Examiner Review Form for HVAC System Design (Loads, Equipment, Ducts)

ACCA Air Conditioning Contractors of America

Form RPER 1.01 8 Mar 10

County, Town, Municipality, Jurisdiction
Header Information

Contractor: GOOD HVAC INC.

Mechanical License #: XYZ 1234

Building Plan #: PM: 2315765

Home Address (Street or Lot#, Block, Subdivision): 1234 Oak Street

REQUIRED ATTACHMENTS[1]	ATTACHED
Manual J1 Form (and supporting worksheets):	Yes ☒ No ☐
or MJ1AE Form[2] (and supporting worksheets):	Yes ☐ No ☐
OEM performance data (heating, cooling, blower):	Yes ☒ No ☐
Manual D Friction Rate Worksheet:	Yes ☒ No ☐
Duct distribution system sketch:	Yes ☒ No ☐

HVAC LOAD CALCULATION (UMC 1106.1)

Design Conditions

Winter Design Conditions

Outdoor temperature	2	°F
Indoor temperature	70	°F
Total heat loss	51,838	Btu

Summer Design Conditions

Outdoor temperature	91	°F
Indoor temperature	75	°F
Grains difference	38 ΔGr @ 50	% Rh
Sensible heat gain	26,107	Btu
Latent heat gain	21,423	Btu
Total heat gain	4,684	Btu

Building Construction Information

Building

Orientation (Front door faces): South
North, East, West, South, Northeast, Northwest, Southeast, Southwest

Number of bedrooms	4
Conditioned floor area	1,600 Sq Ft
Number of occupants	5

Windows

Eave overhang depth	2 Ft
Internal shade (Blinds, drapes, etc)	N/A
Number of skylights	0

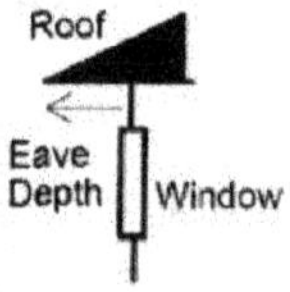

HVAC EQUIPMENT SELECTION

Heating Equipment Data

Equipment type (Furnace, Heat pump, Boiler, etc.)	ABC company
Model	FU600360
Heating output capacity (Heat pumps - capacity at winter design outdoor conditions)	56,000 Btu
Auxiliary heat output capacity	N/A Btu

Cooling Equipment Data

Equipment type (Air Conditioner, Heat pump, etc)	ABC Company
Model	ACO30
Sensible cooling capacity	22,978 Btu
Latent cooling capacity	4,684 Btu
Total cooling capacity	29,298 Btu

Blower Data

Heating CFM	995 CFM
Cooling CFM	995 CFM

HVAC DUCT DISTRIBUTION SYSTEM DESIGN (UMC 601.2)

Design airflow	995 CFM	Longest supply duct:	60 Ft
External Static Pressure (ESP)	0.50 IWC	Longest return duct:	40 Ft
Component Pressure Losses (CPL)	0.20 IWC	**Total Effective Length (TEL)**	233 Ft
Available Static Pressure (ASP) (ASP = ESP - CPL)	0.30 IWC	**Friction Rate:** (Friction Rate = (ASP × 100) ÷ TEL)	0.08 IWC

Duct Materials Used (circle)
Trunk Duct: Duct board, Flex, Sheet metal, Lined sheet metal, Other (specify): Sheet Metal
Branch Duct: Duct board, Flex, Sheet metal, Lined sheet metal, Other (specify): Flex

I declare the load calculation, equipment selection, and duct system design were rigorously performed based on the building plan listed above. I understand the claims made on these forms will be subject to review and verification.

Contractor's Printed Name: Very Good   Date: 1/11/2016

Contractor's Signature: Very Good

Reserved for use by County, Town, Municipality, or Authority having jurisdiction.

[1] The AHJ shall have the discretion to accept Required Attachments printed from approved ACCA software vendors, see list on page 2 of instructions.
[2] If abridged version of Manual J is used for load calculation, then verify residence meets requirements, see Abridged Edition Checklist on page 13 of instructions.

Figure D1: Illustrative Example of a Form for Presenting Information to Code Officials – Furnace and Air Conditioner (Forms available at: http://www.acca.org/standards/codes)

Heat Pump Example

ACCA Air Conditioning Contractors of America

Residential Plans Examiner Review Form for HVAC System Design (Loads, Equipment, Ducts)

Form RPER 1.01
8 Mar 10

County, Town, Municipality, Jurisdiction
Header Information

Contractor GOOD HVAC INC.

Mechanical License # XYZ 1234

Building Plan # PM: 2315767

Home Address (Street or Lot#, Block, Subdivision) 1235 Oak Street

REQUIRED ATTACHMENTS[1]	ATTACHED
Manual J1 Form (and supporting worksheets):	Yes ☒ No ☐
or MJ1AE Form[2] (and supporting worksheets):	Yes ☐ No ☐
OEM performance data (heating, cooling, blower):	Yes ☒ No ☐
Manual D Friction Rate Worksheet:	Yes ☒ No ☐
Duct distribution system sketch:	Yes ☒ No ☐

HVAC LOAD CALCULATION (UMC 1106.1)

Design Conditions

Winter Design Conditions

Outdoor temperature	2	°F
Indoor temperature	70	°F
Total heat loss	51,838	Btu

Summer Design Conditions

Outdoor temperature	91	°F
Indoor temperature	75	°F
Grains difference	38 Δ Gr @ 50	% Rh
Sensible heat gain	26,107	Btu
Latent heat gain	21,423	Btu
Total heat gain	4,684	Btu

Building Construction Information

Building

Orientation (Front door faces) South
North, East, West, South, Northeast, Northwest, Southeast, Southwest

Number of bedrooms 4

Conditioned floor area 1,600 Sq Ft

Number of occupants 5

Windows

Eave overhang depth 2 Ft

Internal shade N/A
Blinds, drapes, etc

Number of skylights 0

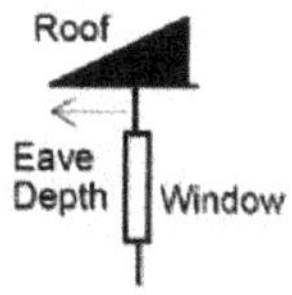

HVAC EQUIPMENT SELECTION

Heating Equipment Data

Equipment type ABC company
Furnace, Heat pump, Boiler, etc.

Model HP5500

Heating output capacity 53,000 Btu
Heat pumps - capacity at winter design outdoor conditions

Auxiliary heat output capacity 34,121 Btu

Cooling Equipment Data

Equipment type ABC Company
Air Conditioner, Heat pump, etc

Model HPCO30

Sensible cooling capacity 22,978 Btu

Latent cooling capacity 4,684 Btu

Total cooling capacity 29,298 Btu

Blower Data

Heating CFM 995 CFM

Cooling CFM 995 CFM

HVAC DUCT DISTRIBUTION SYSTEM DESIGN (UMC 601.2)

Design airflow 995 CFM

External Static Pressure (ESP) 0.50 IWC

Component Pressure Losses (CPL) 0.20 IWC

Available Static Pressure (ASP) 0.30 IWC
ASP = ESP - CPL

Longest supply duct: 60 Ft

Longest return duct: 40 Ft

Total Effective Length (TEL) 233 Ft

Friction Rate: 0.08 IWC
Friction Rate = (ASP × 100) ÷ TEL

Duct Materials Used (circle)

Trunk Duct: Duct board, Flex, Sheet metal, Lined sheet metal, Other (specify)
Sheet Metal

Branch Duct: Duct board, Flex, Sheet metal, Lined sheet metal, Other (specify)
Flex

I declare the load calculation, equipment selection, and duct system design were rigorously performed based on the building plan listed above. I understand the claims made on these forms will be subject to review and verification.

Contractor's Printed Name Very Good Date 1/10/2016

Contractor's Signature Very Good

Reserved for use by County, Town, Municipality, or Authority having jurisdiction.

[1] The AHJ shall have the discretion to accept Required Attachments printed from approved ACCA software vendors, see list on page 2 of instructions.
[2] If abridged version of Manual J is used for load calculation, then verify residence meets requirements, see Abridged Edition Checklist on page 13 of instructions.

Figure D2: Illustrative Example of a Form for Presenting Information to Code Officials – Heat Pump
(Forms available at: http://www.acca.org/standards/codes)

Another set of information that HVAC Contractors may want to present is a completed duct system layout which depicts:

1. Each room's design CFM
2. The duct information
    a. Duct group numbers for: plenums, reducers, and other fittings
    b. Trunk sizes and air flow
    c. Branch sizes
3. Grille and Register Information
    a. Manufacturer
    b. Model
    c. Size
4. Thermostat Information
    a. Manufacturer
    b. Model
    c. Location

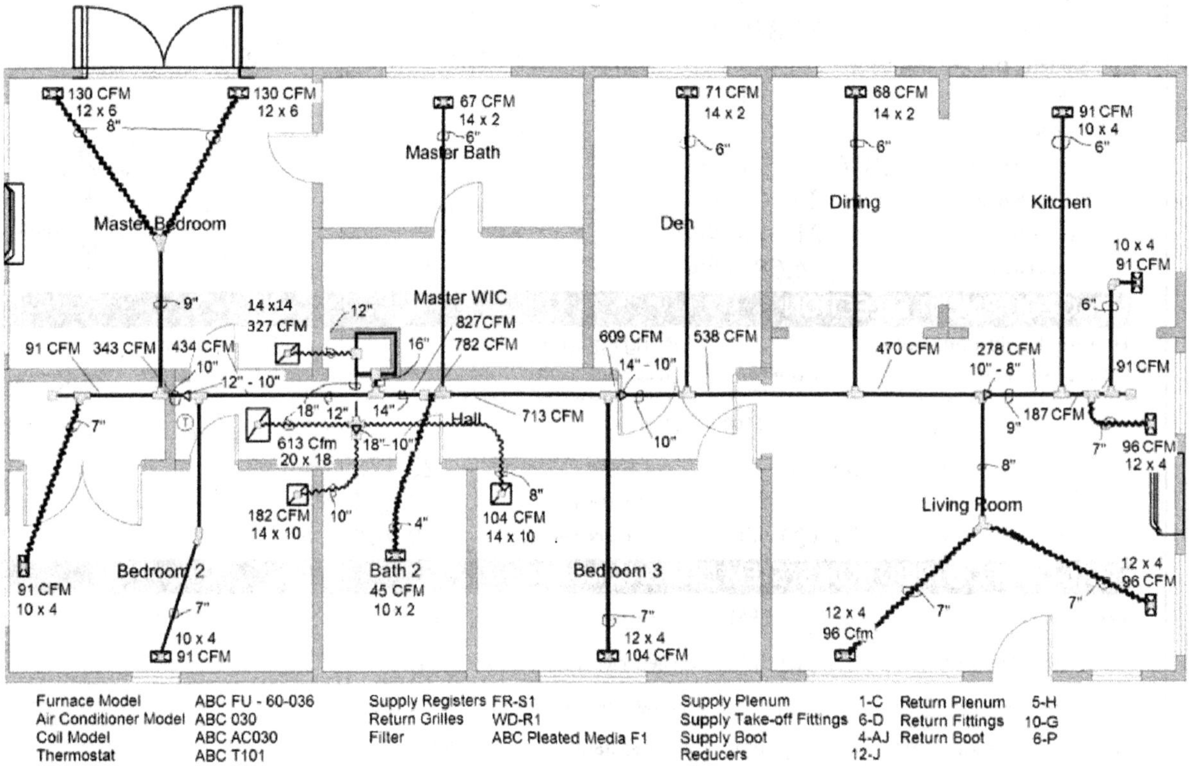

Figure D3: Illustrative Example of Drawing Presenting Information to Code Officials

APPENDIX E
GLOSSARY

ACCA: The Air Conditioning Contractors of America.

AHRI: Abbreviation for Air-Conditioning. Heating and Refrigeration Institute. A non-profit, voluntary organization comprised of heating, air conditioning and refrigeration manufacturers. AHRI publishes standards for testing and rating heat pumps and air conditioners.

air balancing: Adjusting an air conditioning system so that the right amount of air is delivered to the right places in the home in order to achieve the right heating or cooling effect.

air conditioner: Equipment with the ability to absorb heat in occupied spaces and reject it outdoors. The air conditioner may be in one cabinet (package unit) or in two sections (split system). The split system will be a condenser and either a coil or an air handler.

air distribution system: A network of continuous passageways made from tubular or rectangular (thin-wall) pipes and connectors (elbows, tees, branch fittings, and boot fittings), dampers, and plenums used to move air from one point to another.

air handler: An equipment cabinet that contains a blower (and other devices, such as coils and filters) that moves air through the air distribution system. (A forced air furnace provides the same function.)

air makeup: A flow of outdoor air that replaces a flow of mechanically exhausted air (by an engineered system, by infiltration or by a combination of these alternatives).

air outdoor (outside): Air taken from the atmosphere that surrounds the dwelling, which has not been previously circulated in the dwelling, and which is not contaminated by air (or vent gas) that has been exhausted from the dwelling.

air terminal device: A functional decorative cover for air duct. See DIFFUSER, GRILLE, REGISTER

air ventilation: An engineered flow of air drawn from outdoors and introduced to the conditioned space (whether or not processed by the comfort conditioning equipment).

ASHRAE: The American Society of Heating, Refrigeration and Air Conditioning Engineers.

BTU: British Thermal Unit, The amount of heat required to increase or decrease the temperature of one-pound water one degree Fahrenheit.

CFM: Cubic Feet per Minute of air flow: Flow rate for ducted flow, or for the flow through primary heating and cooling equipment, supply outlets, returns and so forth.

Code: Regulations, and subsequent amendments, or any emergency rule or regulation that the administrative authority having jurisdiction has lawfully adopted.

coil: 1) The finned copper or aluminum tube assembly used for heat transfer in refrigerators (evaporator coil), in condensers (condenser coil), and for heat reclaim (heat reclaim coil). 2) Heat exchangers inside an HVAC system that temper and/or dehumidify the air handled by the HVAC system. These include heat exchangers, with or without extended surfaces, through which water, ethylene glycol solution, brine, volatile refrigerant, or steam, is circulated for the purpose of total cooling (sensible cooling plus latent cooling), or sensible heating of a forced-circulation air stream.

commissioning: The final step in installing a heating or air conditioning system. Every component is checked and tested for compliance with codes, good design practices, manufacturer requirements, and occupant needs. Commissioning verifies that the equipment and duct system installation, operation of the equipment, and the performance of the installed system are as intended.

condenser: The outside unit of an air conditioning split system. Here the refrigerant condenses from a gas to a liquid, and hot air from the building is released to the outside. It is also the common term for the outdoor portion of a heat pump split system.

damper: A device for adjusting the amount of air flowing through a duct.

diffuser: A supply air terminal that discharges air in many directions and multiple planes (may, or may not, be equipped with a flow control damper).

draft: The movement of air from an area of high pressure to an area of low pressure; usually considered objectionable. Drafts are also caused by excessive velocity from a supply grille or diffuser.

dry bulb temperature (DB): The temperature, measured by an ordinary thermometer with no evaporative cooling or solar heating effect (the bulb of the instrument is uncovered, dry and in the shade).

duct: Conduits used to carry air. They can be round or rectangular, sheet metal or fiberglass or vinyl tubes. They are also the fittings and connections that link the main airside conduits. In air conditioning it carries air to and from conditioned space or the point of utilization.

ERV: Energy recovery ventilator. Conditions fresh air from outdoors brought into a home while exhausting contaminated air. A special core transfers only sensible (temperature) heat from the exhaust air stream to the incoming air streams or vis-versa, depending on the mode of operation (heating or cooling).

evaporator (heat exchanger): The component of a refrigeration system in which the refrigerant absorbs heat and vaporizes. Usually a coil arranged from pipes or tubing and plates or fins. For air conditioners the evaporator coil is always in the indoor unit. For heat pumps the evaporator is the indoor coil in the summer, and reverses to the outdoor coil in the winter.

fan: A device that produces a pressure difference in air to move it.

fan coil unit (air handler): A cabinet equipped with a blower and one or more coils for heating and/or cooling. The cabinet may accommodate an air filter. The cabinet may have integral supply and return grilles, or it may have inlet and outlet collars for attaching duct runs. When it has outdoor air capability, it may be called a unit ventilator.

Fenestration: Any window, skylight or glass door assembly and associated accessories such as, internal and external shading devices and insect screens.

friction rate: The pressure loss that occurs between two points in a duct system. For duct sizing the formula for friction rate is FR = (ASP × 100) ÷ TEL, where ASP is the available static pressure and TEL is the total effective length.

furnace: That part of a comfort conditioning which converts gas, oil, electricity, or other fossil fuel into heat for distribution within a structure.

geothermal heat pump: A heat pump that absorbs and rejects heat to the ground, sometimes using a well or well water as the heat exchange medium.

grille: A cover for a duct that can make no adjustments to airflow or air pattern. Generally a grille is used on return ducts to let air into the duct system.

heat: Energy which can be transferred as a result of temperature differences.

heat pump: A single refrigeration system designed to provide both heating and cooling. Compare to a furnace and an air conditioner, separate units that only heat or cool.

HRV: Heat Recovery Ventilator. A manufactured mechanical device that reclaims sensible heat, or sensible and latent heat from a flow of exhaust air and transfers the recovered heat to a flow of fresh (outdoor) air.

HVAC: Heating, ventilating, and air conditioning

IAQ: Indoor air quality

infiltration: Air that enters a home through holes, gaps, and cracks, (i.e., plumbing or electrical holes, the heating and air conditioning system, doors, and windows).

latent heat: The heat content (enthalpy) of the air that must be added or removed to cause a change in state (i.e., the heat removed to make ice into liquid water or boiling water to steam). In air conditioning it usually refers to the moisture vapor in the air. Air conditioners remove latent and sensible heat as they cool the air.

load calculation: A mathematical determination of how much cooling and heating (in Btu/h) an HVAC system must deliver for occupant safety and comfort. It is based on a variety of factors: square footage, building orientation, number of occupants, size and placement of rooms, number and size of windows and doors, amount of insulation, number of floors, and climate.

MANUAL D®: An ACCA manual covering the design and installation of residential duct systems.

MANUAL J8®: An ACCA manual covering the method for calculating heating and cooling requirements (load calculation) for single-family detached homes and mobile homes.

MANUAL RS®: An ACCA manual covering the design, installation, and commissioning of a residential HVAC systems.

MANUAL S®: An ACCA manual covering the selection of residential heating and cooling systems.

matched system: A set of compatible HVAC units (i.e., combinations of evaporators, condensers, fan-coils, heat pumps, furnaces, boilers) that have been rated by AHRI, or are in the CEE directory, or are certified by the OEM, to deliver the specified capacity under the design conditions.

mold: A natural byproduct of the fungi family that thrives when organic substances and water combine under certain circumstances. Mold reproduces via spores that can remain dormant, yet viable, for years. Many molds are beneficial. For example, they are the "bleu" in bleu cheese, and we use them to make wine, penicillin, and antibiotics. However, some molds can cause health problems.

NATE: North American Technician Excellence, the nonprofit organization that tests and certifies HVACR technicians.

Occupied space: An enclosed space intended for human activities excluding those spaces intended primarily for other purposes such as storage rooms and equipment rooms that are only occupied occasionally and for short periods of time.

OEM: Original Equipment Manufacturer

PMA: Planned maintenance agreement. Most ACCA member contractors offer PMAs, although they may use different names for them, such as Maintenance Inspection Agreement, Planned Service Agreement, Energy Service Plan, etc.

refrigerant: A fluid that absorbs heat at low temperatures and rejects heat at higher temperatures.

register: An outlet, a grille with a damper assembly, which discharges supply air in a controlled pattern.

relative humidity: The percentage of moisture in the air measured against the amount of moisture the air could hold at a given temperature. Since cold air is capable of holding less moisture than warm air, if the temperature drops and the moisture volume remains constant, the relative humidity will increase. Conversely, if the temperature rises and the moisture volume remains constant, the relative humidity will decrease. If temperature of the air allows it to hold 100 grains of moisture and it is only holding 50 grains, it is at 50% relative humidity.

return, return air, return side: The path the air takes to get to an air-handling unit or furnace so it can be cooled or heated. It is the "return" path. The return side should be "balanced" with the supply side to ensure proper airflow and comfort.

sensible heat: The heat added or removed which causes a change in temperature. In contrast to a heat interchange in which a change of state (latent heat) occurs. Sensible heat can be measured with a thermometer; latent heat cannot.

split system: A two component system with condensing unit installed outside, remote from evaporator section which is installed in conditioned space, uses interconnecting refrigerant lines to connect condensing unit to evaporator. Applies to air conditioners and heat pumps and some refrigeration systems.

spread: The distance an air stream travels horizontally after leaving an air supply outlet, before its force is reduced to a specified terminal velocity.

supply; supply side: The part of an HVAC system that takes (supplies) the conditioned air from the air-handling unit or furnace to the home. The supply side should be "balanced" with the return side to ensure proper airflow and comfort.

temperature: A measure of heat intensity by the heat content of a substance in degrees Fahrenheit or Centigrade

temperature drop (TD) rise (TR): The temperature change between two points in an air distribution system, or water piping system.

temperature, dry-bulb: The temperature of a gas or mixture of gases indicated by an accurate thermometer when there is no heat flow to or from the thermometer bulb.

temperature, wet-bulb: The temperature indicated by a psychrometer, when the bulb is saturated with water over which air is caused to flow at approximately 900fpm to reach an equilibrium temperature of water evaporating into air; when the heat of vaporization is supplied by the sensible heat of the air.

throw: The distance from an air supply outlet that an airstream travels before its force is reduced to a specified terminal velocity.

total effective length (TEL): The total pressure drop that is associated with a duct run. It is equal to the pressure losses that are associated with the straight sections plus the pressure losses that are associated with each and every fitting that is found in the run.

Ventilation: The controlled (engineered) movement of air from the outdoors, through the conditioned space, to the outdoors. The flow of air may exit through cracks and penetrations in the thermal envelope, relief openings, ancillary exhaust systems, dedicated exhaust systems, heat recovery equipment, chimneys or vents.

zones, zoned system, zoning: A single HVAC system controlled by two or more thermostats to meet different heating and cooling needs in different areas (zones). Each zone of a home has its own thermostat with which it can regulate the temperature and humidity in its area. One "zoned air conditioner" could be set for a high temperature in one zone and for a lower temperature in the other zone. Zone systems have two or more zones.

APPENDIX F
Manual S Sizing Limit Overview

Overview of Size Limits for Residential HVAC Equipment

Equipment [a] Tested and Rated by the AHRI	Attributes of Local Climate Notes b, c	Issue	Minimum (deficient) and Maximum(excessive) Capacity Factors. [d]					
		Cooling Capacity (Btuh)	Single-Speed Compressor			Multi- and Variable-Speed Compressor		
			Air-Air	GLHP [e]	GWHP [f]	Air-Air	GLHP [e]	GWHP [f]
Air-Air and Water-Air Cooling-Only & Heat Pump	Mild Winter or Has a Latent Cooling Load	Total	0.90 to 1.15		1.25	0.90 to 1.20 $_{multi}$ or 1.30 $_{variable}$		1.30 $_{m}$ or 1.35 $_{v}$
		Latent	Minimum = 1.00. Preferred maximum = 1.50 (may exceed 1.5 if no reasonable alternative).					
		Sensible	Minimum = 0.90. Maximum determined by total and latent capacities.					
Air-Air and Water-Air Heat Pump Only	Cold Winter and No Latent Cooling load	Total	Maximum capacity = ***Manual J*** total cooling load plus 15,000 Btuh; Minimum factor = 0.90					
		Latent	Latent capacity for summer cooling is not an issue.					
		Sensible	Not an issue (determined by the limits for total cooling capacity).					

a) Central ducted; ductless single-split; ductless multi-split equipment. AHRI: Air Conditioning, Heating and Refrigeration Institute.
b) Mild winter: Heating degree days for base 65°F divided by cooling degree days for base 50°F less than 2.0. Cold winter = 2.0 or more.
c) Latent cooling load: ***Manual J*** sensible load divided by ***Manual J*** total load less than 0.95. No latent load = 0.95 or more.
d) Minimum and maximum capacity factors operate on the total, latent, and sensible capacity values produced by an accurate ***Manual J*** load calculation (per Section 2 of the Eighth Edition of ***Manual J***, version 2.0 or later). Multiply a size factor by 100 to convert to a percentage. For example, 1.15 excess capacity = 115% excess capacity.
e) GLHP: Ground loop heat pump (water in buried closed pipe loop).
f) GWHP: Ground water heat pump (ground water from well, pond, lake, river, etc., flows though equipment and is discarded).

Electric Heating Coils	Furnaces; Heat Pump supplement; emergency	Load (Btuh)	Maximum KW	Minimum Capacity Factor	Maximum Capacity Factor
		≤ 15,000	5.0	Satisfy Load	See Maximum KW
		> 15,000	See Min and Max	0.95	1.75

Minimum and maximum capacity factors operate on the heating load produced by an accurate ***Manual J*** load calculation. Multiply a size factor by 100 to convert to a percentage.

Natural Gas, Oil, Propane Furnaces	Duty	Minimum Output Capacity	Maximum Output Capacity
	Heating-only	1.00	1.40
	Heating-Cooling Preferred		
	Heating-Cooling Allowed		2.00

Minimum and maximum capacity factors operate on the heating load produced by an accurate ***Manual J*** load calculation. Multiply a size factor by 100 to convert to a percentage. For heating-cooling duty, blower performance must be compatible with the cooling equipment.

Electric, and Fossil Fuel Water Boilers	Duty	Minimum Output Capacity	Maximum Output Capacity
	Gravity or forced convection terminals in the space, water coil in duct or air-handler.	1.00	1.40

Minimum and maximum capacity factors operate on the heating load produced by an accurate ***Manual J*** load calculation. Multiply a size factor by 100 to convert to a percentage. Refer to OEM guidance when a boiler is used for potable water heat, or snow melting.

Hot Water Coils	Duty	Minimum Factor	Maximum Factor	
	Gravity or forced convection terminals in the space.	1.00	Two-position 1.25	Throttling 1.50
	Water coil in duct or air-handler.			

Minimum and maximum capacity factors operate on the heating load produced by an accurate ***Manual J*** load calculation. Multiply a size factor by 100 to convert to a percentage. Two-position= open-close valve; Throttling = Full modulating 2-way or 3-way valve.

Electric and Fossil Fuel Water Heaters	The space heating load is the ***Manual J*** load. The total load is the space heating load plus the potable water load. Refer to OEM guidance for selection and sizing guidance.
Dual Fuel Systems	Heat pump sizing rules apply, heating equipment sizing rules apply, see Section N2-11.
Ancillary Dehumidification	See Section N2-12. May allow +15,000 Btuh excess cooling capacity for cold winter climate.
Humidifiers (Section N2-13)	Minimum capacity ≥ humidification load, excess capacity dependent on smallest size available.
AHAM Cooling and Heat Pump Equipment	See Section N2-14 for sizing rules.
Direct Evaporative Cooling Equipment	See Section N2-15 for sizing rules.